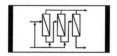

PROTEIN BIOSEPARATION USING ULTRAFILTRATION

Theory, Applications and
New Developments

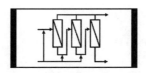

PROTEIN BIOSEPARATION
USING ULTRAFILTRATION
Theory, Applications and
New Developments

Raja Ghosh, D.Phil (Oxon)
McMaster University, Canada

Imperial College Press

ICP

Published by

Imperial College Press
57 Shelton Street
Covent Garden
London WC2H 9HE

Distributed by

World Scientific Publishing Co. Pte. Ltd.
5 Toh Tuck Link, Singapore 596224
USA office: Suite 202, 1060 Main Street, River Edge, NJ 07661
UK office: 57 Shelton Street, Covent Garden, London WC2H 9HE

British Library Cataloguing-in-Publication Data
A catalogue record for this book is available from the British Library.

ISBN   1-86094-317-9

Typeset by Stallion Press
Printed in Singapore.

To my parents

Preface

Proteins have important commercial applications: as pharmaceuticals, food products, food additives, nutraceuticals and industrial catalysts, to name just a few. *Protein bioseparation* refers to the recovery, isolation, purification and polishing of protein products. It is often regarded as the critical limiting factor in the successful commercialisation of protein based products. An ideal protein bioseparation process must combine high productivity with high selectivity of separation, and must be feasible at mild operating conditions. All these requirements are met by *ultrafiltration*, which is a pressure-driven membrane-based separation process.

This book discusses how ultrafiltration could be used for protein bioseparation. There are several good books on protein bioseparation and indeed several others on ultrafiltration. However, there are relatively fewer books dealing specifically with protein bioseparation using ultrafiltration, in spite of this being an area with tremendous potential for growth and development. This book is primarily intended for academic and industrial researchers keen to explore the exciting possibility of using ultrafiltration for protein bioseparation. This book will also be useful for graduate students doing courses in the broad areas of biotechnology and membrane technology. It might be worth mentioning here that the principles and processes discussed in connection to protein bioseparation are also relevant to the bioseparation of other types of biopolymers such as nucleic acids.

From a protein bioseparation point of view, ultrafiltration is mainly used for protein concentration, diafiltration, clarification and fractionation. The first three mentioned here are established technologies, which have already found wide acceptance in the bioprocess industry. Protein fractionation using ultrafiltration is a new, exciting and challenging proposition, and has been dealt with in more detail in this book.

In Chapter 1 of this book, an overview on protein bioseparation is given. Its importance in the overall protein manufacturing process is highlighted and some of the commonly used techniques are briefly discussed. In Chapter 2, ultrafiltration is introduced to the reader. A membrane is the key component of an ultrafiltration process: Chapter 3 deals with membranes. Of almost equal importance is the membrane module, which is the device within which a membrane-based process is carried out. Chapter 4 highlights the role of the membrane module in an ultrafiltration process. Membrane *fouling* is widely regarded as the 'Achilles heel' of ultrafiltration. Chapter 5 discusses this problem. Membrane fouling is also discussed in subsequent chapters in the context of specific phenomena and processes. The productivity of an ultrafiltration process is largely determined by the *permeate flux*. Factors affecting permeate flux along with flux enhancement methods are discussed in Chapter 6. Another important aspect of protein ultrafiltration is the transport of proteins through membranes. This is dealt with in Chapter 7. Chapter 8 very briefly introduces the concept of selectivity. Chapters 9–12 deal with the different types of ultrafiltration based protein bioseparation processes, i.e., protein concentration, diafiltration, clarification and fractionation respectively. New developments and potential areas of further research and development are discussed in Chapter 13.

Thanks are due to several people for their words of encouragement and valuable suggestions regarding the book; in particular, Professors Tony Fane, Richard Bowen, Georges Belfort, Ron Childs and Zhanfeng Cui. My family has been a continuous source of encouragement throughout this endeavour. Special thanks go out to my family members. I would particularly like to acknowledge the contribution made by my wife in terms of encouragement and support, and more specifically, help with the manuscript preparation.

Oxford
March 2002

Acknowledgements

The author acknowledges with thanks the assistance provided by the individuals and organisations named below:

Dr Vikki Chen for permission to reproduce micrographs of membranes (c, d and e of Fig. 3.3) in this book.

J. Dana Hubbard and George Adams of Millipore Corporation for permission to reproduce micrographs of membranes (a and b of Fig. 3.3), photographs of membrane equipment (Figs. 4.2, 4.8 and 4.12) and drawings (Figs. 4.7 and 4.9) in this book.

Markus Simon of Sartorius A/G for permission to reproduce a photograph of membrane equipment (Fig. 4.5) in this book.

Gale Rudd of PCI Membranes for permission to reproduce a photograph of membrane equipment (Fig. 4.10) in this book.

Pall Corporation for permission to reproduce technical data (Fig. 9.2) in this book.

Elsevier Science for permission to reproduce technical data and figures from papers published by the author.

Contents

List of Tables

List of Figures

Chapter 1

Protein Bioseparation: An Overview

1.1 Introduction

Protein bioseparation which refers to the recovery and purification of protein products from various biological feed streams is an important unit operation in the food, pharmaceutical and biotechnological industry. For the purpose of simplicity, these industries will be collectively referred to as *bioprocess industries* throughout this book. Protein bioseparation is at the present moment more important in the bioprocess industry than at any time before. This is largely due to the phenomenal developments in recent years in the field of modern biotechnology. More and more protein products have to be purified in larger quantities. A further boost to protein bioseparation is likely to come from the developing science of *proteomics*.

The purpose of this chapter is to provide the reader with an overview on protein bioseparation. Different aspects of protein bioseparation are discussed in the book edited by Sadana [1]. In order to read about *bioseparations* in general, refer to the book by Belter *et al.* [2].

1.2 Proteins

A protein is a biopolymer composed of basic building blocks called amino acids. Naturally occurring proteins are made up of up to 20 different amino acids. Proteins are by far the most abundant biopolymers in living cells (constituting about 40 to 70 percent of dry cell weight) and have diverse biological functions:

a. Structural components (e.g. collagen, keratin)
b. Catalysts (e.g. enzymes, catalytic antibodies)

c. Transport molecules (e.g. haemoglobin, serum albumin)
d. Regulatory substances (e.g. hormones)
e. Protective molecules (e.g. antibodies)

A protein molecule can be a single poly-(amino acid) chain or may comprise more than one poly-(amino acid) chain, held together by covalent bonds or by non-covalent interactions. A protein usually coils up and folds into a specific 3-dimensional configuration, depending on the intrinsic properties of the protein as well as on the environment in which the protein exists. The structure of a protein can be defined at different levels, these being:

a. Primary
b. Secondary
c. Tertiary
d. Quaternary

The primary structure of a protein is the sequence of amino acids present in the poly-(amino acid) chain/s. The secondary structure describes the local structure of linear segments of the protein molecule. The three most common types of secondary structure are the alpha helices, the beta sheets, and the turns. The tertiary structure is the three-dimensional arrangement of all the atoms present in a single poly-(amino acid) chain. The quaternary structure describes the arrangement of the poly-(amino acid) chains (or subunits) in a particular protein. For details on proteins, refer to the following books [3–5].

1.3   Protein products

As mentioned in the previous section, proteins have a diverse range of biological functions. Proteins also have a diverse array of applications. A large number of protein based products have been commercialised. These can be classified into the following broad categories:

a. Food and nutritional products
b. Pharmaceutical products
c. Industrial catalysts

d. Diagnostic products
e. Proteins used for other miscellaneous applications

Some of the protein-based products are listed in Table 1.1. The first two named categories follow intuitively from the importance of proteins in living systems. A large number of protein products are used as foods, food additives and as nutraceuticals. These are obtained from various microbial, plant and animal sources. Depending on their specific applications, these need to be processed (e.g. purified) to varying degrees. By the rule of thumb, nutraceuticals have greater purity requirements than do food additives and these in turn have to be processed to a greater extent than foods.

Pharmaceutically useful proteins are frequently referred to as *therapeutic proteins*. Most of the recent developments in the area of protein bioseparation are centred on therapeutic proteins.

Enzymes, which are biological catalysts, can be used *in vitro* for industrial scale catalysis. These enzymes are referred to as industrial enzymes and are produced in large quantities. Another major use of enzymes is in diagnostics. Enzymes are also used as components of detergent formulations and cosmetic products.

1.4   The requirement for protein bioseparation

Most protein-based products need to be purified before they can be used. The need for purification is due to the following:

a. Reduction in bulk
b. Concentration enrichment
c. Removal of specific impurities (e.g. toxins from therapeutic products)
d. Prevention of catalysis other than the type desired (as with enzymes)
e. Prevention of catalysis poisoning (as with enzymes)
f. Recommended product specifications (e.g. pharmacopoeial requirement)
g. Enhancement of protein stability
h. Reduction of protein degradation (e.g. by proteolysis)

Table 1.1. Protein products.

Proteins

Food / Food additives / Nutraceuticals
Egg albumin
Casein
Soy proteins
Whey protein concentrate
Protein hydrolysates
Alpha lactalbumin
Beta lactoglobulin
Lysozyme

Pharmaceuticals
Monoclonal antibodies
Serum albumin
Serum immunoglobulins
Factor VIII
Tissue plasminogen activator
Urokinase
Streptokinase
Insulin
Erythropoietin
Alpha and beta interferon
Factor IX

Industrial enzymes
Hemmicellulase
Glucose isomerase
Alpha amylase
Penicillin G acylase
Alkaline proteases
Cellulases

Diagnostic enzymes
Peroxidase
Glucose oxidase

Miscellaneous
Enzymes used in cosmetic products
Detergent enzymes
Digestive enzymes
Enzyme based silage additive

Some of the characteristic features of most protein products are:

a. These products are present at very low concentrations in their respective biological feed streams
b. These products, are present, along with large numbers of impurities, some of which are only slightly different from the products themselves
c. These products are thermolabile
d. These products are sensitive to operating conditions, such as pH and salt concentration
e. These products are sensitive to chemical substances, such as surfactants and solvents
f. The quality requirements for these products are frequently very demanding

These above-mentioned factors imply that an ideal protein bioseparation process must combine high productivity with high selectivity of separation, and must be feasible at *mild* operating conditions.

1.5   Economic aspects of protein bioseparation

The isolation and purification of proteins from the product streams of bioreactors and other biological feed streams is widely recognised to be technically and economically challenging. Protein bioseparation quite often becomes the limiting factor in the successful development of protein based products. The isolation and purification cost can be a substantial fraction of the total cost of production for most products of biological origin. Table 1.2 shows the bioseparation cost as approximate proportion of cost of production for certain protein based products. As clearly indicated by these figures, bioseparation cost is the major cost and this is an incentive for developing cost-effective isolation and purification processes.

1.6   Protein bioseparation methods

A myriad of protein bioseparation techniques is available. Some of these protein isolation and purification techniques are discussed in the following books [1, 6–10]. When developing a bioseparation process for a specific

Table 1.2. Cost of protein bioseparation.

Product	Approximate relative price	Bioseparation cost as % of total cost of production
Food/additives	1	10–30
Nutraceuticals	2–10	30–50
Industrial enzymes	5–10	30–50
Diagnostic enzymes	50–100	50–70
Therapeutic enzymes	50–500	60–80

protein, the following should be taken into consideration:

a. The volume or flow rate of the feed stream
b. The relative abundance of the protein in the feed stream
c. A profile of the impurities present
d. The intended application of the protein, along with particular product specifications
e. The market price of the protein

Protein bioseparation techniques can be classified into three broad categories:

a. High-productivity, low-resolution
b. High-resolution, low-productivity
c. High-resolution, high-productivity

Most conventional protein bioseparation processes rely on a scheme, which is best described as RIPP (Removal, Isolation, Purification and Polishing) [2]. Biological feed streams are generally dilute with respect to the target proteins, which need to be separated from a large number of impurities. Such a feed stream would easily overwhelm a high-resolution separation device. Therefore, low-resolution, high-productivity techniques are used first to reduce the volume and the overall concentration of the process stream. This is followed by high-resolution, low-productivity techniques to obtain the pure target protein. However, with the advent of high-resolution, high-productivity techniques, it is frequently possible to shorten, if not totally replace the RIPP scheme.

Table 1.3 lists some of the more commonly used protein bioseparation techniques. Note that ultrafiltration is listed in two categories since the

Table 1.3. Protein bioseparation techniques.

High-productivity, low-resolution

Cell disruption
Precipitation
Centrifugation
Liquid-liquid extraction
Microfiltration
Ultrafiltration
Supercritical fluid extraction

High-resolution, low-productivity

Ultracentrifugation
Packed bed chromatography
Affinity separation
Electrophoresis
Supercritical fluid chromatography

High-resolution, high-productivity

Fluidised bed chromatography
Membrane chromatography
Ultrafiltration
Monolith column chromatography

resolution in an ultrafiltration process depends very much on how it is operated. Some of the other protein bioseparation techniques are briefly discussed below.

1.6.1 *Cell disruption*

Different types of cells (e.g. microbial, animal and plant) produce proteins either intracellularly or extracellularly. For recovering intracellular proteins, the cells have to be disrupted. Different cell disruption techniques are listed in Table 1.4.

1.6.2 *Precipitation*

Proteins can be partially purified using precipitation techniques. The main advantage of these techniques is that very large process volumes can be handled. Proteins can be precipitated using (a) salting out salts

Table 1.4. Cell disruption methods.

Physical methods

Disruption in ball mill or pebble mill
Disruption using colloid mill
Disruption using French press
Disruption using ultrasonic vibrations

Chemical methods

Disruption using detergents
Disruption using enzymes
Combination of detergent and enzymes
Disruption using solvents

---

like ammonium sulfate and sodium chloride, (b) solvents like ethanol, methanol and acetone, and (c) concentrated acids or alkali. Precipitation processes are generally favoured at low temperatures. After precipitation, the precipitates are separated from the bulk liquid (also called the supernatant) using centrifugation or filtration.

1.6.3   *Centrifugation*

A centrifuge is a device that is used for separating precipitated proteins from a solution by spinning the samples at rotation speeds typically ranging from 1000–10000 revolutions per minute. Centrifugation may be carried out at two different scales:

a. Analytical centrifugation
b. Preparative centrifugation

Analytical centrifuges are used in research laboratories and in the industry for small-scale separation and sample preparations (i.e., 1–1000 ml). Preparative centrifuges handle larger sample volumes (i.e., 1 to several thousand litres).

1.6.4   *Ultracentrifugation*

An ultracentrifuge is a special type of centrifuge, which is operated at a much higher speed, e.g. 30000 revolutions per minute. Ultracentrifuges of both analytical and preparative scales are available. These are used to separate proteins in solution.

1.6.5 *Column chromatography*

Chromatography relies on the distribution of components to be separated between two phases: a stationary (or binding) phase and a mobile phase, which carries these components through the stationary phase. In its simplest form the stationary phase is present in the form of a packed bed within a column, hence the term column chromatography. The mixture of components enters a chromatographic column along with the mobile phase, and each individual component is flushed through the system at a different rate. The rate of migration of a component depends on its interactions with the stationary phase, as well as on the mobile phase flow rate. Different types of columns are used for chromatographic separations. Packed beds are most commonly used. Other types include packed capillary columns, open tubular columns and monolith columns.

Different types of separation chemistries are used for chromatographic separation of proteins:

a. Ion exchange
b. Reverse phase partitioning
c. Hydrophobic interaction
d. Size exclusion
e. Supercritical fluid extraction
f. Affinity interaction

1.6.6 *Electrophoresis*

Electrophoresis refers to the separation of components by employing their electrophoretic mobility (i.e., movement in an electric field). The mixture is added to a conductive medium, followed by the application of an electric field across it. Positively charged components will migrate towards the negative electrode, negatively charged components will migrate towards the positive electrode, while neutral components will remain immobile. Electrophoresis can be classified into two types, depending on the medium in which the separation is carried out:

a. Gel electrophoresis
b. Liquid phase electrophoresis

1.6.7 *Membrane chromatography*

Column chromatography has several major limitations. Some of these limitations could be overcome by using synthetic microporous membranes as chromatographic media [11,12]. In membrane chromatography, the transport of proteins to their binding sites takes place by convection and hence these processes are fast. Thus, the high resolution of a chromatographic process can be combined with the high productivity of a membrane separation process.

1.6.8 *Microfiltration*

Microfiltration relies on the use of microporous membranes for the separation of micron-sized particles from fluids. The various applications of microfiltration include:

a. Cell harvesting from bioreactors
b. Virus removal for solutions
c. Clarification of fruit juice and beverages
d. Water purification
e. Air filtration (for sterilisation)
f. Media sterilisation in bioreactors

References

1. A. Sadana (ed.), *Bioseparation of Proteins* Academic Press, New York (1998).
2. P.A. Belter, C.L. Cussler and W.-S. Hu, *Bioseparations* John Wiley and Sons, New York (1988).
3. T.E. Creighton, *Proteins, 2nd Edition* W.H. Freeman and Company, New York (1993).
4. T.E. Creighton (ed.), *Protein Folding* W.H. Freeman and Company, New York (1992).

5. A. Fersht, *Structure and Mechanism in Protein Science* W.H. Freeman and Company, New York (1999).
6. S. Roe (ed.), *Protein Purification Techniques, 2nd Edition* Oxford University Press, Oxford (2001).
7. R.K. Scopes, *Protein Purification: Principles and Practice* Springer-Verlag, New York (1982).
8. F. Franks (ed.), *Protein Biotechnology* Humana Press, New Jersey (1993).
9. G. Walsh and D.R. Headon, *Protein Biotechnology* John Wiley and Sons, Chichester (1994).
10. S. Doonan (ed.), *Protein Purification Protocols* Humana Press, New Jersey (1996).
11. K.G. Briefs and M.R. Kula, 'Fast protein chromatography on analytical and preparative scale using modified microporous membranes' *Chemical Engineering Science* **47** (1992) 141.
12. D.K. Roper and E.N. Lightfoot, 'Separation of biomolecules using adsorptive membranes' *Journal of Chromatography A* **702** (1995) 3.

Chapter 2

Ultrafiltration: An Overview

2.1 Introduction

Ultrafiltration (UF) is a pressure-driven, separation process in which membranes having pore sizes ranging from 10–1000 Å are used for the concentration, diafiltration, clarification and fractionation of macromolecules (e.g. proteins, nucleic acids, and synthetic polymers) [1–6]. Membrane based separation processes are generally classified on the basis of the membrane pore size or on the type of material being processed (see Fig. 2.1). However, it must be emphasised here that membrane pore size is not the sole basis for separation in ultrafiltration processes. Indeed, as will be discussed in this book, a membrane can be made to retain a 'smaller' molecule while allowing a 'larger' molecule to pass through [7–9]. This phenomenon which is referred to as *reversed selectivity* has aroused considerable interest in using ultrafiltration for macromolecular fractionation processes.

A considerable overlap exists between different types of membrane separation processes and in some cases classification becomes difficult. Generally speaking, ultrafiltration deals with filtration of macromolecules. However, in several applications, smaller molecules or even particulate material are processed by ultrafiltration. A glossary of terms used in membrane based processes is given at the end of this book (see Appendix A).

2.2 Applications of ultrafiltration

UF has a broad variety of applications, ranging from the processing of biological macromolecules to wastewater treatment. Some of the major

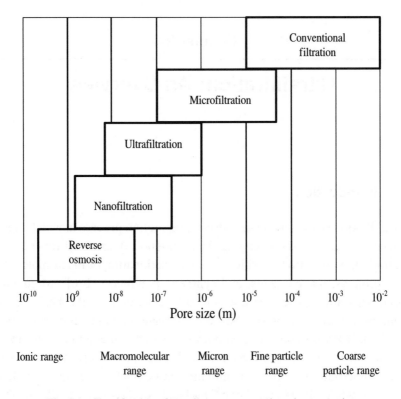

Fig. 2.1. Classification of membrane processes based on pore size.

applications are listed below:

a. Fractionation of macromolecules, e.g. proteins, nucleic acids
b. Concentration of macromolecules, i.e., removal of solvent from solutions of these macromolecules
c. Diafiltration, i.e., removal of salts and other low molecular weight compounds from solutions of macromolecules
d. Removal of cells and cell debris from fermentation broth
e. Virus removal from therapeutic products
f. Harvesting of biomass, e.g. cells and sub-cellular products
g. Membrane reactors
h. Effluent treatment

Food and biotechnological applications account for nearly 40 percent of current total usage of ultrafiltration membranes. Processing of biological macromolecules such as proteins and nucleic acids has assumed significant importance in the bioprocess industry, where the impact of downstream processing on the overall process economics is now being appreciated. In the case of high value therapeutic protein and DNA based products, separation and purification cost can be as high as 80 percent of the total cost of production. Therefore, it makes good economic sense to develop cost-effective and scaleable purification processes for such products. Large-scale protein fractionation is at present more important than large-scale DNA purification since most of the successfully commercialised bioproducts happen to be proteins. However, with progress in the fields of gene therapy and nucleic acid vaccines, there will be a demand for scaleable nucleic acid purification processes.

When it comes to processing proteins, UF is mainly used for:

a. Protein concentration
b. Desalting (or more generally, removal of low molecular weight compounds from protein solution)
c. Clarification (i.e., removal of particles from protein solutions)
d. Protein fractionation (i.e., protein–protein separation)

Concentration, desalting and clarification processes have been widely used in the bioprocess industry for quite some time. Protein fractionation using ultrafiltration is significantly more challenging and is a more recent development.

2.3 Advantages of ultrafiltration

The major advantages of ultrafiltration over competing separation techniques such as chromatography, electrophoresis and affinity separation are:

a. High throughput of product
b. Relative ease of scale-up
c. Ease of equipment cleaning and sanitisation

Techniques traditionally used for protein fractionation in research laboratories (e.g. chromatography, electrophoresis, and affinity separation) are excellently suited for purifying small quantities of proteins. However, these processes are difficult to scale-up, and this limits the scale of production. In addition to scale-up problems, techniques such as chromatography and electrophoresis require complex and expensive equipment. As mentioned earlier, ultrafiltration is widely used for protein diafiltration, clarification and concentration but the potential for its use for protein fractionation is largely untapped [10]. A significant number of ultrafiltration users are unaware of its intricacies and, more importantly, the true potentials of ultrafiltration as a protein separation technique. It is still largely regarded as a simple sieving process where solute (i.e., protein) size is the sole criteria for separation. Early attempts to fractionate proteins purely based on size were unsuccessful. However, solute size is just one of the many factors that could be utilised for separation. Protein–protein interactions, protein-membrane interactions, the extent of concentration polarization and the predominant mode of protein transport (i.e., convective or diffusive) are amongst several factors, which can be exploited for enhancement of protein fractionation. Most membrane researchers have confined their studies to the separation of simulated mixtures of proteins. While this has undoubtedly led to a better understanding of the mechanisms of protein transport and separation, the lack of substantial application based research has kept ultrafiltration in the blind spot of potential users.

References

1. S.S. Kulkarni, E.W. Funk and N.N. Li, 'Introduction and definitions', W.S.W. Ho, K.K. Sirkar (eds.) *Membrane Handbook* Van Nostrand Reinhold, New York (1992) p. 393.
2. M. Cheryan, *Ultrafiltration Handbook* Technomic Pub. Co., Lancaster, Pennsylvania (1986).
3. W.F. Blatt, 'Principles and practice of ultrafiltration', P. Meares (ed.) *Membrane Separation Processes* Elsevier Science Publishing Co., Amsterdam (1976) p. 81.
4. A.S. Michaels, 'Ultrafiltration', E.S. Perry (ed.) *Progress in Separation and Purification*, Vol. 1, Wiley-Interscience, New York (1968) p. 297.

5. E. Flaschel, C. Wandrey and M.R. Kula, 'Ultrafiltration for the separation of biocatalysts', A. Fiechter (ed.) *Advances in Biochemical Engineering/Biotechnology*, Vol. 26, Springer Verlag, Berlin (1983) p. 73.
6. A.G Fane, 'Ultrafiltration: Factors influencing flux and rejection', R.J. Wakeman (ed.) *Progress in Filtration and Separation*, Vol. 4, Elsevier Scientific Publishing Co., Amsterdam (1986).
7. A. Higuchi, S. Mishima and T. Nakagawa, 'Separation of proteins by surface modified polysulfone membranes' *Journal of Membrane Science* **57** (1991) 175.
8. S. Saksena and A.L. Zydney, 'Effect of solution pH and ionic strength on the separation of albumin from immunoglobulins (IgG) by selective filtration' *Biotechnology and Bioengineering* **43** (1994) 960.
9. Q.Y. Li, Z.F. Cui and D.S. Pepper, 'Fractionation of HSA/IgG by gas sparged ultrafiltration' *Journal of Membrane Science* **136** (1997) 181.
10. R. Ghosh and Z.F. Cui, 'Protein purification by ultrafiltration with pre-treated membrane' *Journal of Membrane Science* **167** (2000) 47.

Chapter 3

Membranes

3.1 Introduction

The key component of a membrane separation system is the membrane itself. A membrane is defined as *a thin barrier or film through which solvents and solutes are selectively transported*. An ideal ultrafiltration membrane must have the following characteristics [1]:

a. High hydraulic permeability towards solvent
b. Appropriate sieving property (i.e., selectivity)
c. Good mechanical durability
d. Good chemical and thermal stability
e. Compatibility with substances being processed
f. Excellent manufacturing reproducibility
g. Ease of manufacture

3.2 Membrane material and chemistry

Ultrafiltration membranes are prepared from organic polymers or inorganic material (such as glass, metals and ceramics) [2]:

Organic polymers: Polysulfone (PS), polyethersulfone (PES), cellulose acetate (CA), regenerated cellulose, polyamides (PA), polyvinylidedefluoride (PVDF), polyacrylonitrile (PAN).

Inorganic material: γ-Alumina, α-alumina, borosilicate glass, pyrolyzed carbon, zirconia/stainless steel, zirconia carbon.

19

Figure. 3.1 shows the chemical structures of some of the organic polymeric material used for membrane preparation. Polysulfone, polyethersulfone and cellulose based membranes are more widely used than other types of polymeric membranes. Organic polymer based membranes are more popular than inorganic membranes, due to the following:

a. They are inexpensive
b. They were developed before inorganic membranes
c. They are light
d. They are flexible and can easily be cast or moulded into various shapes and sizes
e. Certain membrane types (e.g. hollow fibres) can only be prepared with organic polymers
f. A wide range of membrane chemistry is available

Inorganic membranes have the following advantages:

a. They can withstand higher transmembrane pressure
b. They are generally more durable (although ceramic and glass membranes can be quite brittle and hence susceptible to breakage)

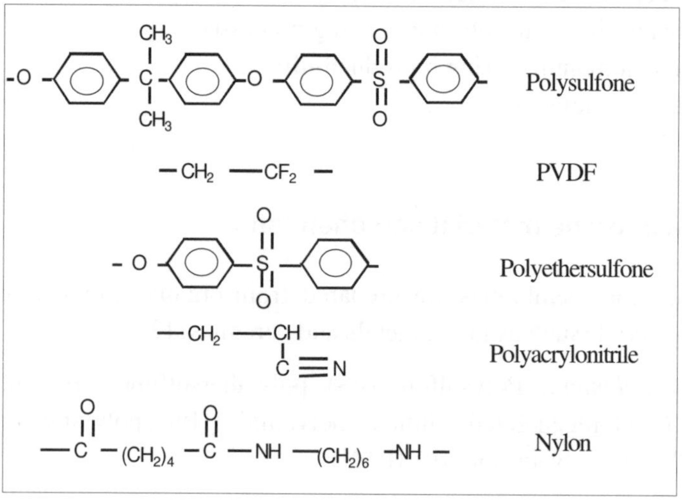

Fig. 3.1. Repeat units of various polymers used for preparing membranes.

c. They are generally resistant to a wider variety of chemical substances, e.g. acids, alkali and solvents

d. They can be easily cleaned and sterilised

Inorganic membranes are more widely used for operations such as water purification and filtration of industrial chemicals.

3.3 Membrane structure and morphology

From a structural point of view, 'membranes can be broadly divided into two types', as shown in Fig. 3.2: (a) symmetric, and (b) asymmetric (or anisotropic). A symmetric membrane has similar structural morphology at all positions within it. An anisotropic membrane is a composite of two or more structural planes of dissimilar morphologies. The selectivity and hydraulic permeability of an anisotropic membrane is usually due

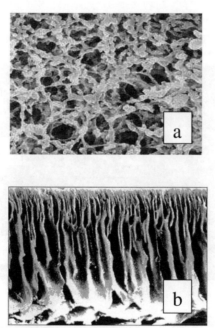

Fig. 3.2. Membrane structure: (a) symmetric and (b) anisotropic [Micrograph (a) courtesy of Millipore Corporation].

to a compact *skin layer*, which can be from 0.2 to 1 micron thick. The skin layer is backed up by a microporous-supporting layer, which gives form and mechanical support. Most polymeric membranes used for protein ultrafiltration are anisotropic. However, some of the newly developed isoporous membranes (i.e., those having all pores of nearly the same diameter) are symmetrical. Most inorganic membranes are symmetrical, though they are rarely used for protein ultrafiltration.

From a morphological point of view, membranes can be porous or dense. A porous membrane has tiny pores or pore networks within it (see Fig. 3.3). On the other hand, dense membranes do not have any pores at all and solute or solvent transport through these membranes takes place by a solubilization-diffusion mechanism. All ultrafiltration membranes are porous. Dense membranes are used for processes such as dialysis and pervaporation.

Fig. 3.3. Pores and pore network in synthetic membranes [Micrographs (a) and (b) courtesy of Millipore Corporation; Micrographs (c), (d) and (e) courtesy of Dr V. Chen].

3.4 Membrane preparation

Most polymeric membranes are prepared by an appropriate casting technique. Polymer casting can be done in four ways [3]:

a. Precipitation from vapour phase: This involves the penetration of a precipitating material from a vapour phase into a homogeneous polymeric film.
b. Precipitation by evaporation: The membrane forming polymer is dissolved in a mixture of a more volatile and a less volatile solvent. As the more volatile solvent is evaporated, the polymer precipitates in the form of a membrane.
c. Immersion precipitation: This involves the immersion of a cast polymeric film in a bath of non-solvent for coagulation leading to membrane formation.
d. Thermal precipitation: This involves the precipitation of a polymer from a solution by cooling.

Other membrane making procedures include [3]:

a. Stretching: This involves the stretching of a polymer film at normal or elevated temperature in order to produce pores of desired size and shape. The pores thus obtained tend to be elongated.
b. Sintering: Powdered material can be sintered by compression with or without heating to give a microporous structure.
c. Slip casting: The method involves coating several layers of uniform particles with decreasing sizes on a porous support. Some inorganic membranes are prepared using this method.
d. Leaching: Some inorganic membranes are prepared by a leaching technique. Isotropic glass membranes are prepared by a combination of phase separation and acid leaching.
e. Track etching: A homogeneous polymeric film is exposed to laser beams or to beams of collimated charged particles. This breaks specific chemical bonds within the polymer matrix. The film is then placed in an etching bath to remove the damaged sections, thus giving rise to monodisperse pores. Isoporous membranes are prepared by track etching.

3.5 Driving force in membrane separation processes

Different driving forces are encountered in membrane separation processes. Some of these are applied to transport solute and solvent molecules through membranes. On the other hand, some (like osmotic pressure) result primarily as a reaction to the applied driving force. The forces include:

a. Transmembrane pressure (TMP)
b. Concentration or electrochemical gradient
c. Osmotic pressure
d. Electrical field

Figure. 3.4 summarizes some of the forces driving solute and solvent transport through an ultrafiltration membrane. The transmembrane pressure (TMP) is the main applied driving force. The TMP for a typical membrane module (shown in Fig. 3.5) is given by the expression:

$$\Delta P = \frac{P_i + P_o}{2} - P_f \qquad (3.1)$$

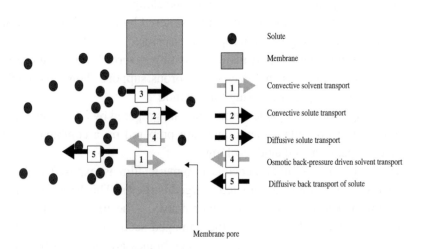

Fig. 3.4. Solute and solvent transport in ultrafiltration.

where

ΔP = Transmembrane pressure (Pa)
P_i = Module inlet pressure (Pa)
P_o = Module outlet pressure (Pa)
P_f = Pressure on the filtrate side (Pa)

The TMP range used in ultrafiltration is 0–500 kPa. Due to the applied TMP, the bulk liquid medium (i.e., the solvent) is forced through the pores. The solvent molecules carry the dispersed phase (i.e., the solute molecules) towards the membrane, and in certain cases through the membrane. Solute molecules may be fully transmitted, partially transmitted or totally retained (or rejected) by the membrane. The retention (or rejection) of solute molecules by the membrane depends on steric, hydrodynamic, thermodynamic and electrostatic effects. The transport of solute molecules through an ultrafiltration membrane takes place largely due to bulk convection (i.e., they are carried along by the solvent).

When a solute is partially retained, there is a build-up of rejected solute molecules near the membrane surface. This leads to a solute concentration difference across the membrane (i.e., the solute concentration of the upstream side is higher than that on the downstream side) resulting in diffusive solute transport through the membrane. However, at reasonably high

$$\Delta P = \frac{P_i + P_o}{2} - P_f$$

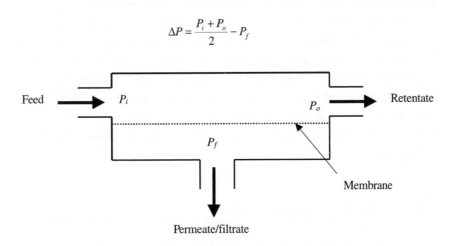

Fig. 3.5. Transmembrane pressure in membrane module.

TMP values the magnitude of diffusive solute transport is small in comparison to convective transport.

The concentration difference across the membrane also leads to the development of a transmembrane osmotic pressure difference, which encourages the flow of solvent from the downstream side back to the upstream side. This is referred to as osmotic back-pressure and acts against the applied TMP. The magnitude of the osmotic back-pressure is a function of the solute concentrations on the upstream and downstream sides of the membrane. This can commonly be expressed in terms of polynomial expressions, as shown below [4, 5]:

$$\Delta\pi = (A_1 C_w^2 + A_2 C_w + A_3) - (A_1 C_p^2 + A_2 C_p + A_3) \qquad (3.2)$$

where

$\Delta\pi$ = Osmotic pressure difference (Pa)

C_w = Solute concentration on the upstream side (kg m^{-3})

C_p = Solute concentration on the downstream side (kg m^{-3})

A_1, A_2 and A_3 are virial coefficients.

The accumulation of solute molecules near the membrane surface also results in a significant concentration difference with respect to the bulk feed, leading to back-diffusion of these accumulated solute molecules. The rate of back-diffusion depends on the transport properties of the solute, as well as on the hydrodynamics of the feed side of the membrane.

Externally applied electrical fields are sometimes used to increase the efficiency of ultrafiltration processes. An electric field is commonly used to encourage the back-transport of accumulated solute molecules from the membrane surface to the bulk feed. An electric field could also be applied to encourage the transport of specific solute molecules through a membrane.

3.6   Membrane characterisation

The performance of a membrane process depends to a great extent on the properties of the membrane itself. Thus membrane characterisation is an important exercise for membrane developers and users. Some of these performance-indicating parameters which need to be determined are:

a. Mechanical strength, e.g. tensile strength, bursting pressure
b. Chemical resistance, e.g. pH range, compatibility with solvents
c. Hydraulic permeability
d. Average porosity and pore size distribution
e. Sieving properties, e.g. nominal molecular weight cut-off
f. Electrical properties, e.g. membrane zeta potential

Ultrafiltration processes are operated in the 0–500 kPa TMP range. A membrane should therefore be strong enough to withstand this applied pressure. The membrane should also be able to withstand shear forces resulting from material flow, as well as vibrations caused by other system components such as pumps. It should also be mechanically durable to withstand reuse, repeated dismantling of the membrane module and cleaning procedures. Mechanical strength and durability are tested by standard techniques used to test polymeric materials. The breaking stress is a parameter commonly determined. Deformation, which may lead to change in membrane sieving properties, can be a serious problem. With inorganic membranes, other specialised materials testing methods are also used.

A membrane has to be compatible with all the chemical substances used during processing, as well as during cleaning. The membrane should be stable over a wide pH range for operational reasons. In certain separation processes, alkaline or acidic pH are found to be suitable. A large range of chemical substances is used for membrane cleaning (e.g. alkali, detergents, and solvents). The membranes should be stable in the presence of these cleaning agents. Membrane manufacturers usually specify chemical substances that may be used to clean specific membranes.

Proteins as such do not represent any challenge toward membranes from the point of view of chemical stability. However, proteins do have a strong tendency to adsorb on different types of surfaces. A good ultrafiltration membrane should be low protein binding. The amount of protein bound is usually measured by static methods, i.e., by soaking a strip of membrane in protein solution and measuring the amount adsorbed using an appropriate analytical technique. However, during protein ultrafiltration, a membrane is usually exposed to significantly elevated protein concentrations, due to the concentration polarisation phenomenon. Therefore, a static binding method gives a comparative indicator at best.

The hydraulic permeability of a membrane is an important property and depends on the percentage porosity, the pore size distribution and the membrane thickness. The higher the hydraulic permeability, the higher the potential productivity. However, the sieving properties of a membrane should not be compromised. The hydraulic permeability is determined by filtering pre-filtered deionised water through the membrane at different TMP. The permeate flux (filtration rate per unit membrane surface area) is plotted against the applied TMP (as shown in Fig. 3.6) and the slope of the straight line thus obtained gives the hydraulic permeability of the membrane (which is expressed as permeate flux per unit applied pressure). If the permeate flux versus TMP plot in deionised water filtration is not linear (i.e., the hydraulic permeability is not constant), the membrane can be assumed to be susceptible to pressure induced deformation.

All pores in a membrane are not necessarily of the same diameter. In most cases, there is a pore-size distribution (see Fig. 3.7). This distribution, which is usually mono-modal, can be sharp or broad. In certain

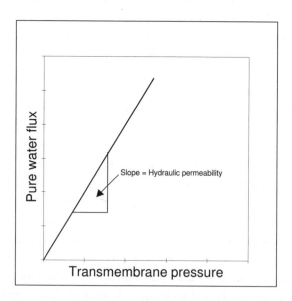

Fig. 3.6. Determination of membrane hydraulic permeability.

Hydraulic Permeab.

cases ... lso encountered. A good ultrafiltration
memb ... ono-modal pore-size distribution. Isoporous ... pared by track etching techniques, have
pores a ... diameter. The membrane pore size can

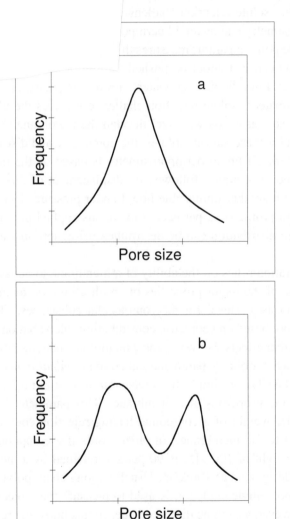

Fig. 3.7. Pore size distribution, frequency versus pore size: (a) mono-modal, and (b) bi-modal.

be measured by direct or by indirect methods. Direct methods involve the use of techniques such as electron microscopy [6,7] or atomic force microscopy (AFM) [8–10]. Both these techniques can provide detailed and magnified 'images' of the membrane from which the pore size and distribution can be determined. Indirect methods include bubble point measurement, solute retention challenge, mercury intrusion porosimetry, thermoporosimetry and bi-liquid permporosimetry.

A bubble point apparatus measures the pressure at which a compressed gas (e.g. air) can just about be pushed through a membrane, which has been immersed in a liquid (e.g. water). From this pressure, the pore size may be estimated. Solute retention challenge involves the ultrafiltration of different solutes of known 'size' through the membrane. From the retention data for these various solutes, the pores' size and distribution may be determined. Bi-liquid permporosimetry is based on the impregnation of membrane by a liquid, followed by displacement of this liquid from the pores with another immiscible liquid under pressure. From the liquid displacement profile (i.e., permeate flux versus applied pressure profile), the pore size distribution can be determined using appropriate numerical methods.

Ultrafiltration relies on the ability of a membrane to act as a selective barrier. One of the major properties of an ultrafiltration membrane is its ability to act as a 'sieve' for macromolecular substances. This sieving property is not based on geometric considerations alone but on a combination of different effects. However, most membrane manufacturers prefer to use a somewhat arbitrary parameter called the *nominal molecular weight cut-off* (NMWCO) or simply the *molecular weight cut-off* (MWCO) to specify sieving properties of a membrane. This parameter is defined as the molecular weight of a compound having rejection coefficient of 0.9 with respect to the membrane. In other words if a compound having a molecular weight of 100 kDa is 90 percent retained by a membrane, its MWCO is designated as 100 kDa. Usually a long chain polymer, such as dextran or polyethylene glycol, is used to measure the molecular weight cut-off. However, as will be discussed later in this book, the retention of a compound does not depend on molecular weight alone. Therefore, the use of MWCO is arbitrary and this value is at best an approximate guide for membrane selection.

Most polymeric membranes have some electrostatic charge on their surface. The magnitude of this charge depends on the operational pH and its effect can be influenced by the presence of charged species in solution (e.g. ions, proteins). The surface charge can also be altered by adsorption of different material. Surface charge can have a significant effect on the transport of charged molecular species through a membrane. Understanding and exploiting the electrical interactions between these molecules and the membrane can significantly enhance the effectiveness of a membrane separation process. The surface charge can be quantified in terms of the *membrane zeta potential* (ζ), which can in turn be determined by measuring the *streaming potential* (ΔU) [11–13].

The *membrane zeta potential* can be calculated using the following equation:

$$\zeta = \lambda\left(\frac{\Delta U}{\Delta P}\right) \tag{3.3}$$

where

ΔP = Applied pressure (Pa)
λ = System dependent constant

References

1. A.S. Michaels, 'Ultrafiltration', E.S. Perry (ed.) *Progress in Separation and Purification*, Vol. 1, Wiley-Interscience, New York (1968) p. 297.
2. S.S. Kulkarni, E.W. Funk and N.N. Li, 'Membranes', W.S.W. Ho and K.K. Sirkar (eds.) *Membrane Handbook* Van Nostrand Reinhold, New York (1992) p. 408.
3. M. Moo-Young (ed.), *Comprehensive Biotechnology*, Vol. 2, Pergamon Press, Oxford (1985).
4. V.L. Vilker, C.K. Colton and K.A. Smith, 'Concentration polarisation in protein ultrafiltration: Part II. Theoretical and experimental study of albumin ultrafiltered in unstirred cell' *AIChE Journal* **27** (1981) 637.
5. V.L. Vilker, C.K. Colton, K.A. Smith and D.L. Green, 'The osmotic pressure of concentrated protein and lipoprotein solutions and its significance to ultrafiltration' *Journal of Membrane Science* **20** (1984) 63.
6. K.J. Kim, A.G. Fane, C.J.D. Fell, T. Suzuki and M.R. Dickinson, 'Quantitative microscopic study of surface characteristics of ultrafiltration membranes' *Journal of Membrane Science* **54** (1990) 89.

7. K.J. Kim, V. Chen and A.G. Fane, 'Some factors determining protein aggregation during ultrafiltration' *Biotechnology and Bioengineering* **42** (1993) 260.
8. W.R. Bowen, N. Hilal, R.W. Lovitt and P.M. Williams, 'Atomic force microscope studies of membranes: Surface pore structures of Cyclopore and Anopore membranes' *Journal of Membrane Science* **110** (1996) 233.
9. W.R. Bowen, N. Hilal, R.W. Lovitt and P.M. Williams, 'Atomic force microscopic studies of membranes: Surface pore structures of Diaflo ultrafiltration membranes' *Journal of Colloid and Interface Science* **180** (1996) 350.
10. W.R. Bowen, N. Hilal, R.W. Lovitt and P.M. Williams, 'Visualisation of an ultrafiltration membrane by non-contact atomic force microscopy at single pore resolution' *Journal of Membrane Science* **110** (1996) 229.
11. M. Nystrom, M. Lindstrom and E. Matthiasson, 'Streaming potential as a tool for characterisation of ultrafiltration membranes' *Colloids and Surfaces* **36** (1989) 297.
12. M. Nystrom, A. Pihlajamaki and N. Ehsani, 'Characterisation of ultrafiltration membranes by simultaneous streaming potential and flux measurements' *Journal of Membrane Science* **87** (1994) 245.
13. N. Le Bolay and A. Ricard, 'Streaming potential in membrane processes: Microfiltration of egg proteins' *Journal of Colloid and Interface Science* **170** (1995) 154.

Chapter 4

Membrane Module and Operation

4.1 Membrane elements and modules

Membrane element refers to the basic form in which a membrane is cast. There are broadly three types of membrane elements:

a. Flat sheet
b. Hollow fibre
c. Tubular

An important component of a membrane separation system is the actual equipment, within which the membrane element is housed. This device is also referred to as the membrane module. The design of the membrane module depends largely on the type of membrane element to be housed. The hydrodynamic conditions within a membrane device depend on the type of membrane element, as well as on the specific module design.

Membrane processes can be classified on the basis of the flow pattern within the membrane device [1]. These are shown in Fig. 4.1. Dead-end ultrafiltration is used only for very small-scale and laboratory applications (i.e., for processing of less than 100 ml of feed solution). Most small, medium and large-scale ultrafiltration processes are carried out in the cross-flow mode. The main advantage of cross-flow ultrafiltration is the minimisation of the accumulation of solute and particulate matter near the membrane surface. The cross-flow arrangement also facilitates recirculation of the retentate stream to the feed tank, followed by its mixing with fresh feed. This gives several operational advantages. Other flow patterns shown in Fig. 4.1 are rarely used in ultrafiltration.

Depending on the specific membrane elements used and on the design, membrane modules can be classified into different types. The most

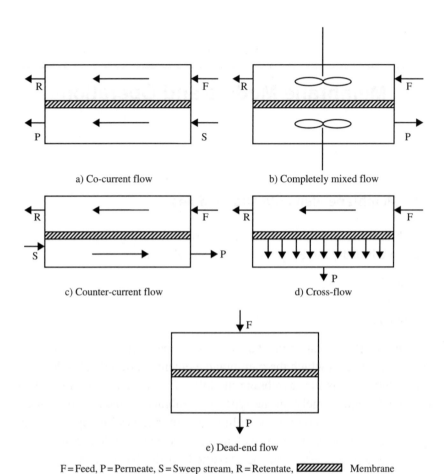

a) Co-current flow

b) Completely mixed flow

c) Counter-current flow

d) Cross-flow

e) Dead-end flow

F = Feed, P = Permeate, S = Sweep stream, R = Retentate, ▨▨▨▨ Membrane

Fig. 4.1. Types of ideal continuous flows in membrane separation processes.

common amongst these are listed below:

a. Stirred cell module
b. Flat sheet tangential flow (TF) module
c. Tubular membrane module
d. Spiral wound membrane module
e. Hollow fibre membrane module

A wide range of membrane modules is now available 'off-the-shelf'. For most membrane applications, the job of the process engineer is to look

Fig. 4.2. Stirred cell ultrafiltration module (Photo courtesy of Millipore Corporation).

at what is available and select the appropriate device for the specific application. However, for certain specialised applications, membrane modules have to be custom designed. This is usually a complicated exercise and involves close co-ordination between the designer, the user and the membrane manufacturer.

Stirred cell modules are useful for small-scale manufacturing and research applications. Figure 4.2 shows an 'off-the-shelf' stirred cell membrane module. This consists of a cylindrical chamber with a disc membrane attached to one of the flat sides (see Fig. 4.3). The content of the stirred cell is kept well stirred using a magnetically driven stirrer bar. Larger stirred cells could have directly driven stirring arrangements. This type of device is commonly used for ultrafiltration and microfiltration processes and is marketed by several companies. Stirred cell modules provide more or less uniform transmembrane pressure and hydrodynamic conditions near the membrane surface. The effects of process parameters on efficiency can be determined easily using stirred cell modules. These are therefore useful for process development work. However, these modules are of not much use in intermediate and large-scale manufacturing operations. Stirred cell devices are generally operated dead-ended. However, as shown in Fig. 4.4, with a continuous feeding and recirculation arrangement, a 'pseudo cross-flow' operation may be carried out [2].

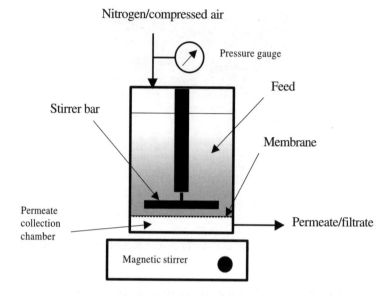

Fig. 4.3. Internal arrangements of a stirred cell ultrafiltration module.

Figure 4.5 shows an 'off-the-shelf' flat-sheet based tangential flow membrane module. The basic flat-sheet tangential flow membrane module consists of a shallow rectangular channel with rectangular flat sheet membrane/s on one or both sides of the channel (see Fig. 4.6). Intermediate and large-scale flat sheet TF modules resemble a plate and frame filter press. The drawing of one such device is shown in Fig. 4.7. These modules consist of alternate layers of membranes, support screens (corrugated or grooved sheets) and distribution chambers for feed, retentate and permeate. These devices can easily be disassembled for cleaning and for replacement of defective membrane elements. Other advantages include the ability to handle reasonably high levels of suspended particulate matter and viscous fluids. Disadvantages include the relatively low membrane packing density. The pressure drop distribution and flow pattern within a TF module having multiple membrane elements in parallel is usually difficult to determine. Consequently, design and analysis of such devices is largely based in empirical methods. A tangential flow filtration module is almost always operated in the cross-flow mode.

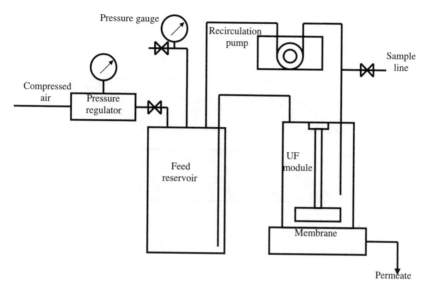

Fig. 4.4. Pseudo cross-flow ultrafiltration using stirred cell module (after Ghosh and Cui [2]).

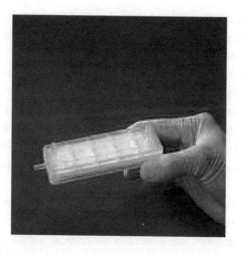

Fig. 4.5. Small-scale tangential flow ultrafiltration unit (Photo courtesy of Sartorius AG, Goettingen).

Figure 4.8 shows a spiral wound membrane module. The module is prepared from flat sheet membranes wound in the form of a spiral envelope

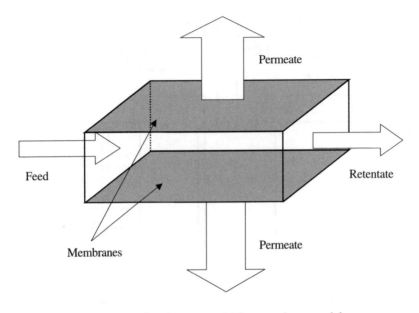

Fig. 4.6. Basic flat sheet tangential flow membrane module.

using a feed spacer (see Fig. 4.9). The feed flows on the outside of the
envelope at high pressure and the permeate is collected on the inside. The
collected permeate runs out of the end of the module. The advantages
of the spiral wound configuration include high membrane packing density
and relatively low cost. Disadvantages include problems with handling
feed containing suspended particulate material, difficulty in cleaning, and
membrane deformation. A major limitation is that these devices cannot
be operated at high transmembrane pressure. For ultrafiltration processes,
a spiral wound membrane module is almost always operated in the cross-
flow mode.

Figure 4.10 shows the section of a tubular membrane module. As the
name indicates, this type of module uses tubular membrane elements (see
Fig. 4.11). These tubes are generally more than 3 mm in diameter. Normally, a tubular membrane module is made up of several tubes arranged as
in a shell and tube heat exchanger. The feed stream enters the lumen of the
tubes (i.e., the inside of the tubes) and the retentate exits from the other end
of the tubes. The permeate passes through the wall (i.e., the membrane)
and is collected on the shell side. The advantages of a tubular membrane

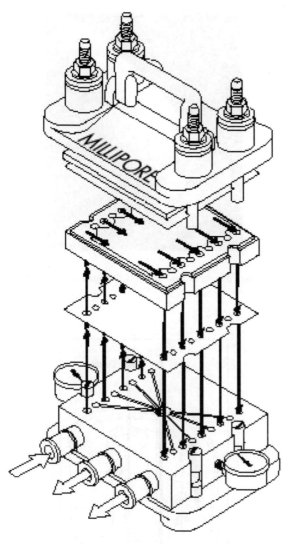

Fig. 4.7. Large-scale flat sheet tangential flow membrane module (Drawing courtesy of Millipore Corporation).

module include turbulent flow (leading to low solute/particulate matter build-up), relatively easy cleaning, easy handling of feed containing suspended particulate matter and viscous fluids, and the ability to replace or plug a failed membrane element. Disadvantages include high capital cost,

Fig. 4.8. Spiral wound membrane module (Photo courtesy of Millipore Corporation).

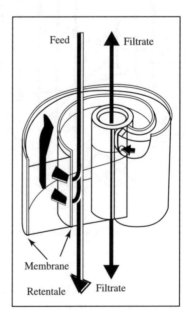

Fig. 4.9. Internal arrangements of a spiral wound membrane module (Drawing courtesy of Millipore Corporation).

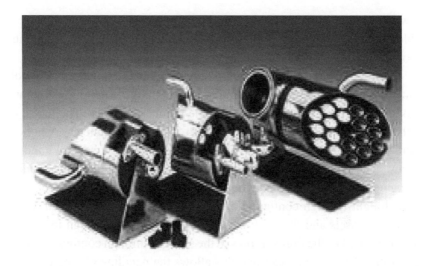

Fig. 4.10. Sections of a tubular membrane module (Photo courtesy of PCI Membranes).

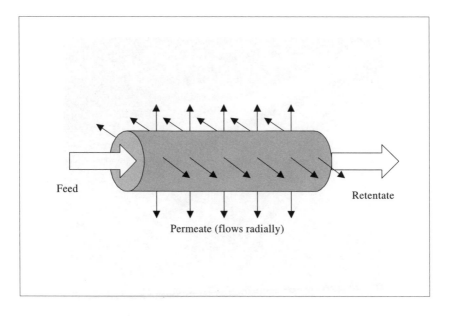

Fig. 4.11. Tubular membrane.

low membrane packing density, high pumping costs, and limited achievable concentrations (due to high hold-up volume). A tubular membrane module is always operated in the cross-flow mode.

Figure 4.12 shows various hollow fibre membrane modules. The hollow fibre membrane module is similar in design to the tubular membrane module, except in terms of scale and the number of tubes. The tubes, or rather hollow fibres, are typically 0.25 to 2.5 mm in diameter. A hollow fibre membrane module usually consists of a bundle of several hundred fibres. These are spun separately, bundled, and potted into tube headers using epoxy resin. These may be bundled together in either a U-shape or in a straight-through configuration. The fibre bundles are housed inside pressure vessels and the feed material normally flows through the inside (or the lumen) of the fibres. Fibres in the straight-through configuration are generally of larger diameter and this allows the handling of feed containing reasonable levels of suspended particulate matter. These substances can easily block the finer fibres in the U-shape configuration. U-shape design is generally used for reverse osmosis, while the straight-through design is used for ultrafiltration. Advantages of hollow fibres include low

Fig. 4.12. Hollow fibre membrane modules (Photo courtesy of Millipore Corporation).

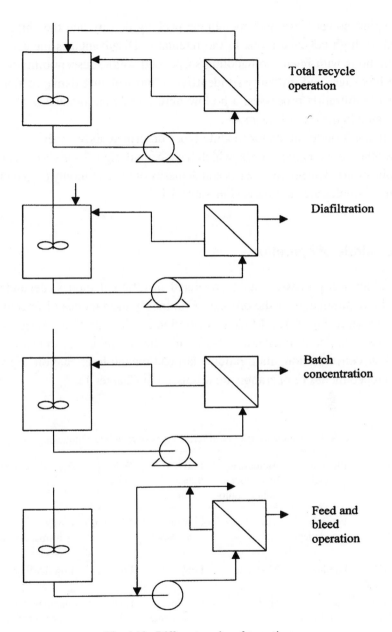

Total recycle
operation

Diafiltration

Batch
concentration

Feed and
bleed
operation

Fig. 4.13. Different modes of operation.

pumping power, very high membrane packing density and the ability to achieve high concentrations in the retentate. Disadvantages include the limitations with regard to handling suspended solids, susceptibility to fibre blockage, and difficulty in identifying and isolating damaged fibres. For ultrafiltration processes, a hollow fibre membrane module is always operated in the cross-flow mode.

In addition to the major module types described above, there are several other less common types, which have been designed for very specific applications. The relative merits and demerits of the commonly used membrane modules are summarised in Table 4.1.

4.2 Mode of operation

Ultrafiltration processes can be operated in different modes, depending on the requirements of the process. Commonly used modes of operation are shown in Fig. 4.13. Mode of operation in ultrafiltration processes is discussed in detail in Chapters 9–12 with specific reference to processes such as concentration, diafiltration, clarification and fractionation. Newly developed modes of operation are discussed in Chapter 13.

Table 4.1. Characteristics of different types of membrane modules.

Type	Fluid flow regime	Membrane area/module volume ratio	Mass transfer coefficient	Hold-up volume	Special remarks
TF flat sheet	Laminar-turbulent	Low	Low to moderate	Moderate	Easily dismantled and cleaned
Spiral wound	Laminar	Moderate	Low	Low	Low TMP only
Hollow fibre	Laminar-turbulent	High	Low to moderate	Low	Susceptible to fibre blocking
Tubular	Turbulent	Low	Moderate to high	Moderate to high	Flow pattern easy to characterise

References

1. Terminology for membranes and membrane processes (IUPAC, 1996) website: http://www.che.utexas.edu/nams/NAMSHP.html (accessed 14 December 2001).
2. R. Ghosh and Z.F. Cui, 'Fractionation of BSA and lysozyme using ultrafiltration: Effect of pH and membrane pretreatment' *Journal of Membrane Science* **139** (1998) 17.

Chapter 5

Membrane Fouling

5.1 Introduction

Fouling refers to the adsorption and deposition of material present in the feed on the membrane leading to loss of efficiency of the separation process. Membrane fouling results in:

a. Reduction in permeate flux due to decrease in hydraulic permeability
b. Alteration in solute transmission behaviour

In protein ultrafiltration, the protein itself is the major foulant. Fouling can be reversible or irreversible in nature. Reversible fouling refers to that type, the effects of which may be reversed by appropriate cleaning procedures. However, some irreversible fouling may also occur, which over time may necessitate membrane replacement. It must be noted that fouling is not the only reason for decrease in permeate flux with time. With certain types of membranes, membrane compaction (i.e., deformation) may also result in flux decline. Sometimes these effects are difficult to distinguish. Specific issues concerning protein-based fouling are discussed in subsequent chapters with reference to processes such as protein concentration, diafiltration, clarification and fractionation.

Fouling is an undesirable effect and a lot of attention has been devoted to prevention of fouling or to put it more pragmatically, minimisation of fouling. This statement is basically an admission that membrane fouling will always take place, no matter how much effort is made to prevent it. However, by carefully analysing factors responsible for fouling and by devising means by which the effects of these can be minimised, fouling can be kept at an acceptably low level.

Several factors influence membrane fouling. These are:

a. Physicochemical properties of the membrane
b. Physicochemical properties of the protein
c. Membrane morphology
d. The operating parameters, e.g. TMP, permeate flux, system hydrodynamics
e. The physicochemical properties of the feed solution, e.g. pH, salt concentration
f. Membrane operation history

5.2   Fouling mechanisms

Fouling by proteins can be due to adsorption or deposition on the external membrane surface, or due to adsorption or deposition within the pores [1]. A protein may get adsorbed on the membrane surface due to electrostatic adsorption, hydrophobic interaction or van der Waals forces. Deposition can take place due to impaction and subsequent loss of momentum. Figure 5.1 shows the different overall mechanisms by which membrane fouling takes place.

5.2.1   *External fouling*

When the adsorption/deposition takes place on the external surface, the hydraulic permeability and solute transmission characteristics are altered due to:

a. Increase in the effective membrane thickness
b. Blockage of pore entrance
c. Constriction of pore entrance

The two main mechanisms by which protein molecules are transported from the feed solution for adsorption or deposition on the external membrane surface are:

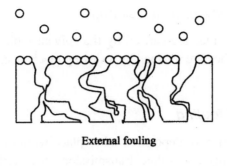

External fouling

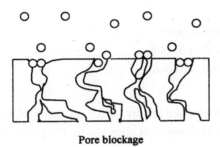

Pore blockage

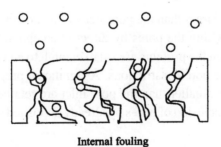

Internal fouling

Fig. 5.1. Overall mechanisms of membrane fouling.

5.2.1.1 *Diffusional transport to membrane*

The protein molecules are transported from the bulk fluid to the membrane surface through the fluid boundary layer by diffusion.

5.2.2 *Internal fouling*

When the adsorption or deposition takes place within the pores, the hydraulic permeability and solute transmission characteristics are altered due to:

a. Internal blockage of pore
b. Internal constriction of pore
c. Alteration of pore tortuosity

There are three mechanisms by which protein molecules find their way to their adsorption or deposition sites within the pores. These are:

5.2.2.1 *Direct interception*

If the proteins are larger than the pore size of the membrane, these proteins will be retained within the pores by direct interception. This mechanism predominates when the proteins are small enough to enter the pores but are still larger than the pore constrictions. A significant proportion of proteins whose diameter is smaller than the pore inlet are retained by bridging effects and partial occlusion of pores by already deposited protein molecules.

5.2.2.2 *Inertial impaction*

The protein molecules in a fluid stream have mass and velocity, and thereby possess momentum. As the carrier fluid meanders along the tortuous pores, taking the path of least resistance to flow, bulkier proteins will tend to continue in the previously established flow direction, impact upon pore wall, lose their momentum and thus get deposited or adsorbed. This mechanism is observed with larger proteins and protein aggregates.

5.2.2.3 *Diffusional transport to pore walls*

This mechanism predominates when the protein molecules are significantly smaller than the pores. The protein molecules are transported from their fluid streamlines to the pore walls by diffusion and thus get adsorbed or deposited.

5.3 Chemistry of adsorption

The adsorption of protein molecules on the membrane surface takes place primarily due to electrostatic and hydrophobic interactions. Proteins are charged molecules and quite frequently membranes themselves carry surface electrostatic charge. The magnitude and type of charge (i.e., positive or negative) depend on the physicochemical conditions. Protein membrane electrostatic interactions will also be influenced by other ionic species present in the feed solution.

Proteins are made up of different types of amino acids, some of which are hydrophobic. In solution, the hydrophobic amino acid residues, in their effort to hide away from water molecules, cluster towards the interior of the protein structure. When a protein molecule comes in close proximity to a hydrophobic surface (as have some of the membranes), the hydrophobic amino acids come out of hiding and interact with the surface. Such interaction is referred to as hydrophobic interaction. Hydrophobic interactions are influenced by the presence of salts in the feed solution.

The exact mechanism of adsorption can be quite complex, with more than one type of interaction frequently involved. The molecular interactions between proteins and synthetic membranes have been examined by Pincet *et al.* [2]. These studies were carried out using a surface force apparatus. The interactions examined include those between proteins and the membrane and those between the membrane-bound proteins and those in solution. Protein membrane interactions were shown to disrupt the structure of the bound proteins. Such disruption was suggested as being responsible for extensive membrane fouling.

The adsorption of proteins on the membrane surface is generally assumed to follow Langmuir type isotherm. However, at elevated protein concentrations, such as those due to concentration polarisation, the

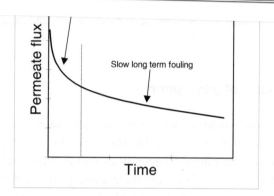

Fig. 5.2. Decline in permeate flux with time in ultrafiltration processes operated at fixed TMP.

Langmuir type isotherm may not hold good any more. At severe concentration polarisation, a gel layer is frequently formed and this may cause more severe permeate flux decline than that by adsorbed protein molecules.

5.4 Initial and long-term fouling

In constant pressure ultrafiltration, membrane fouling is observed in the form of permeate flux decline with time (see Fig. 5.2). There is an initial rapid decline in permeate flux, which takes place within the first 10 to 15 minutes of operation. The decline in permeate flux over the remaining part of the operation is more gradual and less severe. The initial sharp decline in permeate flux is due to the combined effects of concentration polarisation and rapid initial fouling. Rapid fouling takes place particularly when a fresh membrane surface is exposed to proteins. The adsorption sites on the membrane surface are rapidly occupied by protein molecules and this leads to sharp decline in membrane hydraulic permeability. If the same membrane is used for a subsequent ultrafiltration process using the same solution, the initial rapid fouling is less extensive. Thus, the history of a membrane can have a significant effect on fouling.

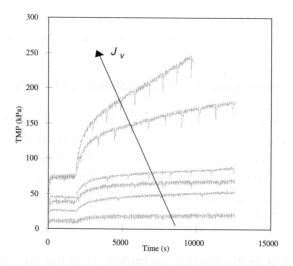

Fig. 5.3. Increase in transmembrane pressure in constant flux ultrafiltration.

Long-term fouling is observed after the first 10 to 15 minutes of operation and is generally less severe. This is usually caused by adsorption of secondary layers of protein and by the deposition of protein molecules within pores already constricted by the rapid initial adsorption.

The effects of fouling on permeate flux have mainly been observed by experiments carried out at constant transmembrane pressure (TMP). This is primarily due to the fact that most ultrafiltration processes are carried out at constant TMP. Some researchers have found it convenient to examine fouling at constant permeate flux conditions, since this ensures constant convective transport of solute and solvent towards the membrane, e.g. [3–7]. In a constant permeate flux operation, the effects of fouling and concentration polarisation are observed in terms of increase in TMP. Ghosh [7] examined the effects of constant flux on increase in transmembrane pressure in totally retentive BSA ultrafiltration (see Fig. 5.3). An initial rapid increase in the transmembrane pressure was observed. This is due to the combined effects of rapid initial fouling and concentration polarisation. After this initial rapid increase, the transmembrane pressure increased linearly and less rapidly over the rest of the filtration process. This was thought to be due to long-term fouling. These studies were carried out at different constant permeate flux values. The extent of initial

5.5 Effect of concentration po...

Control of concentration polarisation can help in reducing fouling, since build-up of protein molecules is found to promote rapid and extensive adsorption, eventually leading to fouling. If the membrane surface concentration (C_w) reaches the point where protein precipitates or forms a thixotropic gel, this gel layer can provide an additional resistance in series with the membrane.

5.6 Effect of permeate flux on fouling, and the critical flux concept

As shown in Fig. 5.3, the extent of fouling increases as the permeate flux is increased. However, below a certain system dependent permeate flux, the extent of fouling is low and is independent of the permeate flux. This is referred to as the *critical flux* [8]. The interrelation of permeate flux with fouling is discussed in the next chapter.

5.7 Effect of physicochemical parameters on fouling

Environmental conditions, such as pH and salt concentration, have also been found to have a profound influence on the rate and extent of membrane fouling for a given protein-solvent system, e.g. [9–11]. This will be discussed in detail in subsequent chapters.

References

1. A.G. Fane, 'Ultrafiltration: Factors influencing flux and rejection', R.J. Wakeman (ed.), *Progress in Filtration and Separation*, Vol. 4, Elsevier Scientific Publishing Co., Amsterdam (1986).
2. F. Pincet, E. Perez and G. Belfort, 'Molecular interactions between proteins and synthetic membrane polymer films' *Langmuir* **11** (1995) 1229.

3. V. Chen, 'Performance of partially permeable microfiltration membranes under low fouling conditions' *Journal of Membrane Science* **147** (1998) 265.
4. A.D. Marshall, P.A. Munro and G. Tragardh, 'Influence of permeate flux on fouling during the microfiltration of beta-lactoglobulin solutions under cross-flow conditions' *Journal of Membrane Science* **130** (1997) 23.
5. P. Aimar, J.A. Howell, M.J. Clifton and V. Sanchez, 'Concentration polarisation buildup in hollow fibres — A method of measurement and its modeling in ultrafiltration' *Journal of Membrane Science* **59** (1991) 81.
6. J.A. Howell, 'Sub-critical flux operation' *Journal of Membrane Science* **107** (1995) 165.
7. R. Ghosh, 'Study of membrane fouling by BSA using pulsed injection technique' *Journal of Membrane Science* **195** (2002) 115.
8. P. Bacchin, P. Aimar and V. Sanchez, 'Model for colloidal fouling of membranes' *AIChE Journal* **41** (1995) 368.
9. K.J. Kim, V. Chen and A.G. Fane, 'Some factors determining protein aggregation during ultrafiltration' *Biotechnology and Bioengineering* **42** (1993) 260.
10. K.L. Jones and C.R. O'Melia, 'Protein and humic acid adsorption onto hydrophilic membrane surfaces: Effects of pH and ionic strength' *Journal of Membrane Science* **165** (2000) 31.
11. M.K. Ko, J.L. Pellegrino, R. Nassimbene and P. Marko, 'Characterisation of adsorption-fouling layer using globular proteins on ultrafiltration membranes' *Journal of Membrane Science* **76** (1993) 101.

Chapter 6

Permeate Flux in Ultrafiltration

6.1 Permeate flux

The volumetric permeate flux J_ν (which is the filtration rate per unit membrane surface area and is usually expressed in m s^{-1}) represents the productivity of a membrane separation process. It depends on the properties of the membrane, the transmembrane pressure, the system hydrodynamics, the protein concentration in the feed and the properties of the solvent and the protein.

The transmembrane pressure for a typical membrane module is given by:

$$\Delta P = \frac{P_i + P_o}{2} - P_f \qquad (6.1)$$

P_i and P_o are the inlet and outlet pressures respectively on the feed side and P_f is the pressure on the filtrate side.

For the filtration of a pure solvent, or a solution having negligibly low protein concentration, the permeate flux is given by the 'pore flow model' [1]:

$$J_\nu = \frac{\varepsilon_m d_p^2 \Delta P}{32 \mu l_p} \qquad (6.2)$$

where
$$\begin{aligned}
\varepsilon_m &= \text{membrane porosity (dimensionless)} \\
d_p &= \text{average pore diameter (m)} \\
\Delta P &= \text{transmembrane pressure (Pa)} \\
\mu &= \text{viscosity (kg m}^{-1}\text{ s}^{-1}) \\
l_p &= \text{average pore length (m)}
\end{aligned}$$

$$J_v = \frac{\overline{}}{R_m}$$

(6.5)

Where

R_m = membrane resistance (Pa s m^{-1})

It is often observed that the permeate flux in an ultrafiltration process does not increase linearly with transmembrane pressure beyond a certain point (see Fig. 6.1). This is due to the build-up of rejected protein

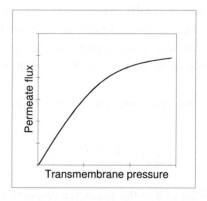

Fig. 6.1. Effect of transmembrane pressure on permeate flux.

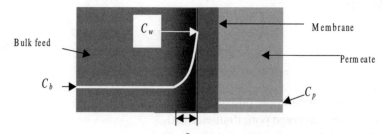

δ_b = Thickness of concentration polarization layer

Fig. 6.2. Concentration polarisation.

molecules near the membrane surface and this phenomenon is referred to as concentration polarisation (see Fig. 6.2). The permeate flux can be represented by the 'concentration polarisation model' [2]:

$$J_\nu = k \ln \left(\frac{C_w - C_p}{C_b - C_p} \right) \qquad (6.4)$$

where

C_b = feed concentration (kg m^{-3})
C_p = permeate (or filtrate) concentration (kg m^{-3})
C_w = wall concentration (kg m^{-3})
k = mass transfer coefficient (m s^{-1}) = (D/δ_b)
D = diffusivity of protein (m^2 s^{-1})
δ_b = boundary layer thickness on the feed side (m)

The concentration polarisation model is based on the assumption that close to the membrane there is a *stagnant film* of thickness δ_b, which is parallel to the membrane surface. This so-called stagnant film corresponds to the hydrodynamic boundary layer. Within this stagnant film, the transport of solvent molecules takes place by convection. The transport of protein molecules towards the membrane takes place by convection while their back-transport away from the membrane towards the bulk feed takes place by diffusion. The mass transfer coefficient, which is essentially a measure of the diffusive back-transport of the protein, depends on the thickness of this stagnant film and on the diffusivity of the protein molecule. Some of the limitations of the concentration polarisation model are due to:

a. Assumption of a stagnant film
b. Assumption of uniform mass transfer coefficient within the stagnant film
c. Assumption of constant protein diffusivity within the film

When the protein being ultrafiltered is totally retained (or rejected) by the membrane, the concentration polarisation equation reduces to:

$$J_\nu = k \ln \left(\frac{C_w}{C_b} \right) \qquad (6.5)$$

The rejected protein molecules, which accumulate near the membrane surface, result in a transmembrane concentration difference. This in turn

resistance model', which takes into consideration the contribution of the osmotic pressure, is given below:

$$J_\nu = \frac{\Delta P - \Delta \pi}{R_m + R_{cp}}$$ (6.6)

where

$\Delta \pi$ = osmotic pressure difference (Pa)
R_{cp} = hydraulic resistance of the concentration polarisation layer (Pa s m^{-1})

At very high permeate flux, the wall concentration of the protein may increase to such an extent that gel layer is formed (see Fig. 6.3). When a protein gel layer is formed, the concentration polarisation equation may be modified to obtain the 'gel polarisation equation':

$$J_\nu = k \ln \left(\frac{C_g}{C_b} \right)$$ (6.7)

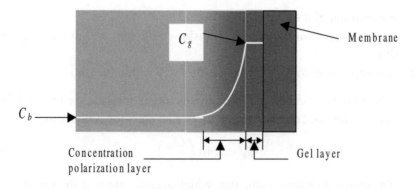

Fig. 6.3. Gel layer.

where

C_g = gelation concentration of the protein (kg m^{-3})

The corresponding hydraulic resistance model now assumes the form:

$$J_\nu = \frac{\Delta P - \Delta \pi}{R_m + R_{cp} + R_g} \tag{6.8}$$

where

R_g = hydraulic resistance offered by the gel layer (Pa s m^{-1})

As shown in Fig. 6.1, the permeate flux does not increase linearly at higher transmembrane pressures. As the pressure is increased even further, the permeate flux levels off (as shown in Fig. 6.4). A still further increase in transmembrane pressure may even lead to a decline in permeate flux. The pressure range in which the permeate flux increases with increase in transmembrane pressure is referred to as the pressure dependent region. The range where the increase in pressure does not increase the permeate flux is referred to as the pressure independent region. The permeate flux in the pressure independent region is referred to as the 'limiting flux'.

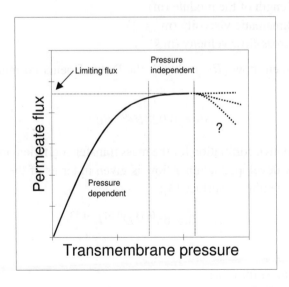

Fig. 6.4. Pressure dependence of permeate flux, and limiting flux concept.

modules having tubular geometry, this parameter may be estimated using dimensionless correlations based on heat-mass transfer analogy. The mass transfer coefficient can be estimated from correlations for the Sherwood number in terms of the Reynolds number and the Schmidt number.

In the case of fully developed laminar flow, the Graetz-Leveque correlation can be used [1]:

$$Sh = 1.62 \, Re^{0.33} Sc^{0.33} \left(\frac{d}{l_t}\right)^{0.33} \tag{6.9}$$

where

Sh = Sherwood number (dimensionless) = (kd/D)
Sc = Schmidt number (dimensionless) = (ν/D)
D = diffusivity of protein ($m^2 \, s^{-1}$)
Re = Reynolds number (dimensionless) = (du/ν)
d = module diameter (m)
l_t = length of the module (m)
ν = kinematic viscosity ($m^2 \, s^{-1}$)
u = cross-flow velocity ($m \, s^{-1}$)

For turbulent flow ($Re > 2000$), the Dittus-Boelter correlation can be used [1]:

$$Sh = 0.023 Re^{0.8} Sc^{0.33} \tag{6.10}$$

An alternative correlation for the mass transfer coefficient in the special case of fully developed laminar flow is given in terms of the shear rate γ (s^{-1}) at the membrane surface [3]:

$$k = 0.816 \gamma^{0.33} D^{0.67} l_t^{-0.33} \tag{6.11}$$

where

γ = $(8u/d)$ for tubes
γ = $(6u/b)$ for rectangular channels (b = channel depth, m)

From the correlations mentioned above, it is clear that increasing the Reynolds number increases the mass transfer coefficient. However, the mass transfer coefficient is more sensitive to the Reynolds number in the turbulent region (see Fig. 6.5). By increasing the cross-flow velocity within the membrane module, the mass transfer coefficient and hence the permeate flux can be increased.

6.2   Enhancement of permeate flux

At a particular transmembrane pressure for a given protein concentration in the feed, the permeate flux depends on the mass transfer coefficient; the higher the crossflow velocity (and therefore the higher the value of k), the higher the permeate flux (see Fig. 6.6). The permeate flux can also be increased by periodically disrupting the concentration polarisation layer. Some of the flux enhancement methods are:

a. Increasing the feed flow rate
b. Creating pulsatile feed flow
c. Pressure pulsing

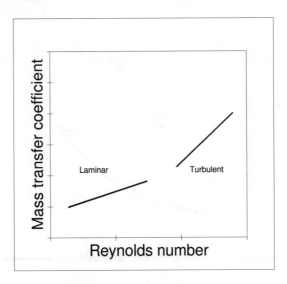

Fig. 6.5. Dependence of the mass transfer coefficient on Reynolds number.

h. Gas-sparging
i. Externally applied electrical field

With 'off-the-shelf' membrane equipment, the permeate flux can be in-creased by increasing the cross-flow velocity. Figure 6.6 shows the effect of cross-flow velocity on permeate flux. This increase is due to a decrease in concentration polarisation within the *stagnant film* as the cross-flow ve-locity is increased. In the laminar flow regime, the mass transfer coefficient does not increase as significantly with cross-flow velocity as it does in the turbulent flow regime. Enhancement of permeate flux by increasing the cross-flow rate is limited by certain constraints. The pump being used usu-ally determines the upper limit for flow rate. The flow rate may also be limited by the pressure drop across the module, which can be very high at high cross-flow rates. In any case, increasing permeate flux by increasing the cross-flow velocity is not always the most cost effective approach.

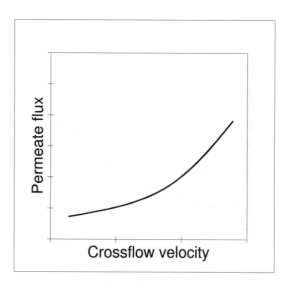

Fig. 6.6. Effect of cross-flow velocity on permeate flux.

The use of pulsatile flow of the feed for flux enhancement has been reported by many researchers, e.g. [4–7]. The time averaged permeate flux is higher with pulsatile flow than with continuous flow (as shown in Fig. 6.7). The effect of pulsatile flow on the permeate flux is thought to be due to the combined effects of induced backflushing and shear-rate changes. Backflushing opens up pores that have somehow been blocked by protein molecules and other material. The pulsatile nature of the flow and consequent shear rate changes also inhibit the development of the concentration polarisation layer.

Pressure pulsing methods have been used for increasing the permeate flux in membrane filtration processes, e.g. [8–12]. This involves the application of timed pressure pulses on the permeate side in order to periodically change the direction of material flow within the pores. Pulsing of the transmembrane pressure reduces membrane fouling resistance and alters the concentration polarisation boundary layer. Pressure pulsing also causes membrane movement, which is thought to contribute towards the disruption of concentration polarisation.

The use of oscillatory flow of feed to enhance permeate flux in membrane filtration processes has been examined by several workers, e.g. [13–17]. The permeate flux enhancement due to oscillatory flow is thought to

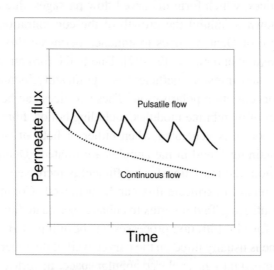

Fig. 6.7. Enhancement of permeate flux using pulsatile flow.

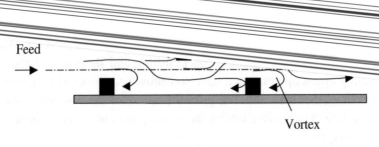

Feed

Vortex

Fig. 6.8. Enhancement of permeate flux by baffles and inserts.

be caused by decreased membrane fouling and intermittent disruption of the concentration polarisation layer.

The use of baffles and tube inserts for creating turbulence and thereby increasing permeate flux has been reported by several workers, e.g. [4, 6, 14, 18–22]. Baffles and tube inserts promote good local mixing of the feed, particularly in regions close to the membrane (see Fig. 6.8) and minimise concentration polarisation effects by increasing the mass transfer coefficient.

Dean vortices, which form in curved flow passages, due to flow instability, are known to inhibit the growth of the concentration polarisation layer. The use of Dean vortices to enhance permeate flux has been reported by several workers, e.g. [23–27]. Due to the formation of the Dean vortices, the mass transfer coefficient is significantly increased, thereby decreasing concentration polarisation. These vortices can be generated in different types of membrane modules, including hollow fibres.

A Taylor vortex is generated when either or both of two concentric cylinders containing liquid in the annulus are rotated. Due to the Taylor vortex formation, the mass transfer coefficient is increased. Thus, significant enhancement in permeate flux can be achieved. Commercial membrane devices utilising Taylor vortex to enhance permeate flux are available and their use for ultrafiltration processes has been reported, e.g. [28–30]. The membrane is usually fitted on the curved wall of the inner cylinder and the feed is pumped into the enclosed annular space. In vortex flow devices, the wall shear rate can be de-coupled from the feed flow rate.

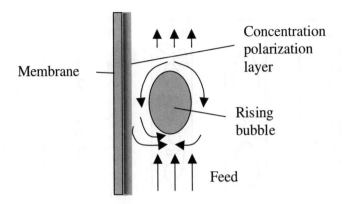

Fig. 6.9. Bubble induced secondary flow.

Gas sparging is widely recognised as an effective way to increase per-meate flux in ultrafiltration processes, e.g. [31–37]. This involves injection of gas (e.g. air, nitrogen) in the form of bubbles into the feed stream. The injected gas bubbles enter the membrane module and improve the hydro-dynamics, mainly due to bubble induced secondary flow (see Fig. 6.9) and increased bulk flow. These effects result in an increase in the mass transfer coefficient, leading to enhancement of permeate flux. Gas sparg-ing has been found to be most effective in tubular membrane modules. It was found to be significantly less effective in hollow fibre membranes. With flat sheet tangential flow devices, gas-sparging was found to be rea-sonably effective from the flux enhancement point of view. With tubu-lar membranes, different two-phase flow (i.e., gas-sparging) regimes were investigated (see Fig. 6.10). Amongst these, the slug flow regime was found to be most effective.

The application of an electric field across a membrane can be an ef-fective means of concentration polarisation control. The use of a pulsed electric field has been found to be more effective than using a constant electric field. A pulsed electric field was found to be very effective in dead-end ultrafiltration of bovine serum albumin [38]. Optimisation of the operating variables, such as the magnitude of the electric field, the pulse in-terval, the pulse duration, and the feed conditions (pH, ionic strength, and protein concentration) was found to be important. The flux enhancement

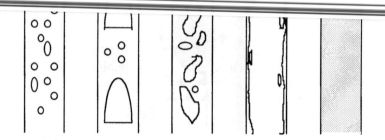

Fig. 6.10. Flow regimes in gas liquid two-phase flow (after Ghosh and Cui [36]).

effect could be explained in terms of the contributions of electrophoretic, electroosmotic, and hydrodynamic forces.

6.3 Fouling control

As mentioned earlier, fouling is a major cause of permeate flux decline in ultrafiltration processes. Fouling in protein ultrafiltration is influenced by the following factors:

a. Permeate flux
b. Concentration polarisation
c. pH
d. Salt concentration
e. Feed concentration
f. Protein aggregation
g. Protein denaturation

Below a certain system dependent permeate flux value the extent of fouling is found to be negligible. This is referred to as the 'critical flux' [39]. Above the critical flux value, the permeate flux has been shown to have a strong influence on fouling, i.e., the extent of fouling increases with increase in permeate flux. In order to avoid problems associated with fouling, it might be better to operate at a sub-critical flux value. However, this might mean a decrease in productivity. In practice a compromise permeate flux value might have to be used to take into consideration these conflicting factors.

The concentration polarisation which depends on both the permeate flux and the mass transfer coefficient has been found to influence the extent of membrane fouling [40]. Concentration polarisation induces fouling by increasing the amount of protein adsorption. The higher the degree of concentration polarisation, the higher the protein concentration near the membrane and consequently the higher the amount of protein adsorbed on the membrane. When the concentration polarisation is severe the wall protein concentration can reach a value called the 'gelation concentration' at which a protein gel layer can form on the membrane surface. Therefore, techniques useful in minimising concentration polarisation would also be useful in minimising membrane fouling.

The feed solution pH and its electrolyte concentration are found to strongly influence membrane fouling in protein ultrafiltration [41]. Both these parameters affect protein–protein and protein-membrane interactions. Protein-membrane interactions influence the fouling behaviour primarily in the initial stages of ultrafiltration. On the other hand, protein–protein interactions influence fouling during the entire operation. Generally speaking, the fouling is more extensive when the protein in the feed solution is uncharged (i.e., at its pI). At low electrolyte (salt) concentration, a greater extent of fouling is observed. As the salt concentration is increased, the extent of fouling decreases. However, at very high salt concentrations, fouling increases again. This increase in fouling is thought to be due to protein precipitation and enhanced hydrophobic interactions between protein and membrane. In order to minimise permeate flux decline due to fouling, the pH and salt concentration of the feed need to be adjusted taking all the above mentioned factors into consideration.

The feed concentration influences fouling by affecting the amount of protein adsorption on the membrane. The amount of protein adsorbed primarily depends on the wall protein concentration. Therefore the effect of feed concentration is convoluted with those of permeate flux and mass transfer coefficient (and therefore concentration polarisation). Wherever possible, the feed should be appropriately diluted to minimise permeate flux decline due to fouling.

tein aggregation results in extensive accumulation of aggregated near the surface (due to a decrease in the effective mass transfer coefficient). Aggregation essentially leads to the transition from concentration polarisation to 'cake formation'. The formation of a stagnant cake having low hydraulic permeability near the membrane surface, due to consolidation of aggregated protein, results in severe permeate flux decline. When a cake layer is formed, concentration polarisation minimising methods are no longer effective. Therefore solution conditions have to be maintained such that protein aggregation is kept at a minimum.

Proteins can be denatured during ultrafiltration due to a combination of various effects. These include the geometry of the membrane module, the pumping mechanism, the physicochemical properties of the feed (i.e., pH and salt concentration), the membrane surface chemistry, the permeate flux and the cross-flow velocity. Denaturation of protein during processing is undesirable and represents a loss in productivity. A denatured protein also causes more severe membrane fouling [43]. Therefore there is an added incentive to ensure that protein denaturation before or during ultrafiltration is kept to a minimum.

References

1. E. Flaschel, C. Wandrey and M.R. Kula, 'Ultrafiltration for the separation of biocatalysts', A. Fiechter (ed.) *Advances in Biochemical Engineering/Biotechnology*, Vol. 26, Springer Verlag, Berlin (1983) p. 73.
2. W.F. Blatt, 'Principles and practice of ultrafiltration', P. Meares (ed.) *Membrane Separation Processes* Elsevier Science Publishing Co., Amsterdam (1976) p. 81.
3. S.S. Kulkarni, E.W. Funk and N.N. Li, 'Theory and mechanistic concepts', W.S.W. Ho and K.K. Sirkar (eds.) *Membrane Handbook* Van Nostrand Reinhold, New York (1992) p. 398.
4. S.M. Finnigan and J.A. Howell, 'The effect of pulsatile flow on ultrafiltration fluxes in a baffled tubular membrane system' *Chemical Engineering Research & Design* **67** (1989) 278.

5. L. Ding, C. Charcosset and M.Y. Jaffrin, 'Albumin recovery enhancement in membrane plasma fractionation using pulsatile flow' *International Journal of Artificial Organs* **14** (1991) 61.

6. Y.Y. Wang, J.A. Howell, R.W. Field and D.X. Wu, 'Simulation of cross-flow filtration for baffled tubular channels and pulsatile flow' *Journal of Membrane Science* **95** (1994) 243.

7. H.Y. Li, C.D. Bertram and D.E. Wiley, 'Mechanisms by which pulsatile flow affects cross-flow microfiltration' *AIChE Journal* **44** (1998) 1950.

8. V.G.J. Rodgers and R.E. Sparks, 'Reduction in membrane fouling in the ultrafiltration of binary protein mixtures' *AIChE Journal* **37** (1991) 1517.

9. V.G.J. Rodgers and R.E. Sparks, 'Effect of transmembrane pressure pulsing on concentration polarisation' *Journal of Membrane Science* **68** (1992) 149.

10. V.G.J. Rodgers and R.E. Sparks, 'Effects of solution properties on polarisation redevelopment and flux in pressure pulsed ultrafiltration' *Journal of Membrane Science* **78** (1993) 163.

11. C. Wilharm and V.G.J. Rodgers, 'Significance of duration and amplitude in transmembrane pressure pulsed ultrafiltration of binary protein mixtures' *Journal of Membrane Science* **121** (1996) 217.

12. J.A. Levesley and M. Hoare, 'The effect of high frequency backflushing on the microfiltration of yeast homogenate suspensions for the recovery of soluble proteins' *Journal of Membrane Science* **158** (1999) 29.

13. J.A. Levesley and B.J. Bellhouse, 'Particulate separation using inertial lift forces' *Chemical Engineering Science* **48** (1993) 3657.

14. M.R. Mackley and N.E. Sherman, 'Cross-flow filtration with and without cake formation' *Chemical Engineering Science* **49** (1994) 171.

15. J.A. Levesley and B.J. Bellhouse, 'The retention and suspension of particles in a fluid using oscillatory flow' *Chemical Engineering Research and Design* **75** (1997) 288.

16. K. Abel, 'Influence of oscillatory flows on protein ultrafiltration' *Journal of Membrane Science* **133** (1997) 39.

17. P. Blanpain-Avet, N. Doubrovine, C. Lafforgue and M. Lalande, 'The effect of oscillatory flow on crossflow microfiltration of beer in a tubular mineral membrane system — Membrane fouling resistance decrease and energetic considerations' *Journal of Membrane Science* **152** (1999) 151.

18. S.M. Finnigan and J.A. Howell, 'The effect of pulsed flow on ultrafiltration fluxes in a baffled tubular membrane system' *Desalination* **79** (1990) 181.

19. V. Mavrov, N.D. Nikolov, M.A. Islam and J.D. Nikolova, 'An investigation on the configuration of inserts in tubular ultrafiltration modules to control concentration polarisation' *Journal of Membrane Science* **75** (1992) 197.

20. S. Elmaleh and N. Ghaffor, 'Cross-flow ultrafiltration of hydrocarbon and biological solid mixed suspensions' *Journal of Membrane Science* **118** (1996) 111.

cation Technology 22 (2001) 59.

23. M.E. Brewster, K.Y. Chung and G. Belfort, 'Dean vortices with wall flux in a curved channel membrane system. 1. A new approach to membrane module design' *Journal of Membrane Science* **81** (1993) 127.

24. K.Y. Chung, R. Bates and G. Belfort, 'Dean vortices with wall flux in a curved channel membrane system. 4. Effect of vortices on permeation fluxes of suspensions in microporous membrane' *Journal of Membrane Science* **81** (1993) 139.

25. K.Y. Chung, W.A. Edelstein and G. Belfort, 'Dean vortices with wall flux in a curved channel membrane system. 6. 2-dimensional magnetic-resonance-imaging of the velocity-field in a curved impermeable slit' *Journal of Membrane Science* **81** (1993) 151.

26. G. Belfort, R.H. Davis and A.L. Zydney, 'The behaviour of suspensions and macromolecular solutions in cross-flow microfiltration' *Journal of Membrane Science* **96** (1994) 1.

27. H.R. Millward, B.J. Bellhouse and G. Walker, 'Screw-thread flow promoters — An experimental-study of ultrafiltration and microfiltration performance' *Journal of Membrane Science* **106** (1995) 269.

28. M. Balakrishnan, G.P. Agarwal and C.L. Cooney, 'Study of protein transmission through ultrafiltration membranes' *Journal of Membrane Science* **85** (1993) 111.

29. M. Mateus and J.M.S. Cabral, 'Modelling membrane filtration of protein and cell-suspensions in a vortex flow filtration system' *AIChE Journal* **41** (1995) 764.

30. M. Balakrishnan and G.P. Agarwal, 'Protein fractionation in a vortex flow filter. I: Effect of system hydrodynamics and solution environment on single protein transmission' *Journal of Membrane Science* **112** (1996) 47.

31. Z.F. Cui and K.I.T. Wright, 'Gas-liquid 2-phase cross-flow ultrafiltration of BSA and dextran solutions' *Journal of Membrane Science* **90** (1994) 183.

32. Z.F. Cui and K.I.T. Wright, 'Flux enhancements with gas sparging in downwards crossflow ultrafiltration: Performance and mechanism' *Journal of Membrane Science* **117** (1996) 109.

33. S.R. Bellara, Z.F. Cui and D.S. Pepper, 'Gas sparging to enhance permeate flux in ultrafiltration using hollow fibre membranes' *Journal of Membrane Science* **121** (1996) 175.

34. Z.F. Cui, S.R. Bellara and P. Homewood, 'Airlift crossflow membrane filtration — A feasibility study with dextran ultrafiltration' *Journal of Membrane Science* **128** (1997) 83.

35. Q.Y. Li, Z.F. Cui and D.S. Pepper, 'Effect of bubble size and frequency on the permeate flux of gas sparged ultrafiltration with tubular membranes' *Chemical Engineering Journal* **67** (1997) 71.

36. R. Ghosh and Z.F. Cui, 'Mass transfer in gas-sparged ultrafiltration: Upward slug flow in tubular membranes' *Journal of Membrane Science* **162** (1999) 91.

37. H.W. Sur and Z.F. Cui, 'Experimental study on the enhancement of yeast microfiltration with gas sparging' *Journal of Chemical Technology and Biotechnology* **76** (2001) 477.

38. W.R. Bowen and A.L. Ahmad, 'Pulsed electrophoretic filter-cake release in dead-end membrane processes' *Chemical Engineering Science* **43** (1997) 959.

39. P. Bacchin, P. Aimar and V. Sanchez, 'Model for colloidal fouling of membranes' *AIChE Journal* **41** (1995) 368.

40. P. Aimar, J.A. Howell, M.J. Clifton and V. Sanchez, 'Concentration polarisation buildup in hollow fibres — A method of measurement and its modelling in ultrafiltration' *Journal of Membrane Science* **59** (1991) 81.

41. I.H. Huisman, P. Pradanos and A. Hernandez, 'The effect of protein–protein and protein-membrane interactions on membrane fouling in ultrafiltration' *Journal of Membrane Science* **179** (2000) 79.

42. V. Chen, A.G. Fane, S. Madaeni and I.G. Wenten, 'Particle deposition during membrane filtration of colloids: Transition between concentration polarisation and cake formation' *Journal of Membrane Science* **125** (1997) 109.

43. M. Meireles, P. Aimar and V. Sanchez, 'Albumin denaturation during ultrafiltration — Effects of operating-conditions and consequences on membrane fouling' *Biotechnology and Bioengineering* **38** (1991) 528.

Chapter 7

Protein Transmission Through Ultrafiltration Membranes

7.1 Protein size

Proteins are polymeric molecules and do not differ significantly in size, even when they differ significantly in molecular weight. By the rule of thumb, an 8- to 10-fold difference in molecular weight is needed to ensure a two-fold difference in Stokes-Einstein radius. From Stokes-Einstein equation, the diffusivity is given by:

$$D = \frac{K_1}{r} \tag{7.1}$$

Where K_1 is a constant and r is the Stokes-Einstein radius. Again, from a typical correlation of diffusivity and molecular weight:

$$D = \frac{K_2}{M^{1/3}} \tag{7.2}$$

Where K_2 is a constant and M is the molecular weight. Therefore

$$r = \frac{K_1}{K_2} M^{1/3} \tag{7.3}$$

This clearly suggests that a 2-fold difference in Stokes-Einstein radius can be ensured only by an 8-fold difference in molecular weight.

7.2 Protein transmission

The first significant attempt to model solute transport through ultrafiltration membranes was by Ferry [1]. The apparent rejection of the solute was

where

R_a = apparent rejection coefficient (dimensionless) = (C_p/C_b)
C_p = permeate concentration (kg m^{-3})
C_b = bulk feed concentration (kg m^{-3})
λ = ratio of solute to pore diameter (dimensionless) = (d_s/d_p)
d_s = solute diameter (m)
d_p = pore diameter (m)

The transport of proteins through ultrafiltration membranes is essentially the transport of colloidal particles through liquid filled pores. For the purpose of a simplified analysis, a protein can be considered to be a rigid sphere and the pore to be a hollow liquid filled cylinder. In protein ultrafiltration, the dimensions of the rigid sphere and the cylinder are of the same order of magnitude. Therefore, the transport of the sphere though the cylinder is expected to be hindered. If factors causing hindrance and their magnitude could be determined, the transport of a protein molecule though a porous membrane could be explained and possibly modelled. A large number of researchers have tried to explain membrane transport of proteins from this 'hindered transport modelling' approach. Deen [2] has reviewed the research work done in this area. Early work in this area was based primarily on analysis of the hydrodynamic hindrances, i.e., hindrance to convection and diffusion. However, with charged molecules and charged pores (as is frequently the case in protein ultrafiltration) electrostatic effects are also expected to be significant. Smith and Deen [3] considered proteins to be charged colloidal particles and examined the effect of the 'electrostatic double layer' on protein transport through regular cylindrical pores.

Bowen and Jenner [4] have given a rigorous, dynamic mathematical model for the ultrafiltration of charged colloidal particles. The scope of this model could easily be extended to explain protein ultrafiltration. This model took into account the electrostatic double layer interactions and was based on the numerical solution of the non-linear Poisson-Boltzmann equation. Bowen and Sharif [5] have presented another mathematical model, based on the solution of the non-linear Poisson-Boltzmann equation, to account for the electrostatic interactions, and the Navier-Stokes equation to account for the hydrodynamic interactions. This model focuses on the hydrodynamic and electrostatic interactions experienced by a charged colloidal particle, as it is about to enter a membrane pore. This emphasis is based on the assumption that the entry of the colloidal particle into the pore is the *main event* and whatever happens afterwards is of secondary interest. More recently, Bowen and Sharif [6] have discussed how the electrostatic interactions of a charged particle with the charged walls of a membrane pore are influenced by the presence of other charged particles in close proximity. Prior to this report, such effects were generally not taken into consideration for modelling membrane transport.

The model developed by Kedem and Katchalsky [7] for solute transport through biological membranes is also useful for modelling protein transport in ultrafiltration. This model takes into account the coupling of solute transport with solvent transport.

$$J_v = L_P(\Delta P - \sigma_d \Delta \pi) \tag{7.6}$$

$$J_S = P\Delta C + J_v(1 - \sigma_f)\overline{C} \tag{7.7}$$

where

L_P = hydraulic permeability (m Pa^{-1} s^{-1})
σ_d = osmotic reflection coefficient (dimensionless)
J_S = solute flux (kg m^{-2} s^{-1})
P = solute permeability (m s^{-1})
ΔC = concentration difference (kg m^{-3})
σ_f = solvent drag reflection coefficient (dimensionless)
$\overline{C}$ = average concentration in the layer (kg m^{-3})

The Kedem-Katchalsky formulation is difficult to use when there is concentration polarisation. The 'concentration polarisation model' (used to explain permeate flux in ultrafiltration processes) can be combined with other transport equations and utilised to model protein transmission through membranes. The membranes used for protein ultrafiltration generally have pores that are larger than the protein molecules. The transmission of a protein through such a membrane has been found to depend on intrinsic or true sieving coefficient (S_i).

$$S_i = \frac{C_p}{C_w} \tag{7.8}$$

S_i depends on the protein-solvent system, the membrane, the physic-ochemical conditions (i.e., pH, salt concentration) and the permeate flux. From diffusive-convective membrane transport theory [8]:

$$S_i = \frac{S_\infty \exp\left(\frac{S_\infty J_v \delta_m}{D_{\text{eff}}}\right)}{S_\infty + \exp\left(\frac{S_\infty J_v \delta_m}{D_{\text{eff}}}\right) - 1} \tag{7.9}$$

where

S_∞ = asymptotic sieving coefficient (dimensionless)
δ_m = membrane thickness (m)
D_{eff} = effective diffusivity of the protein within the membrane (m^2 s^{-1})

As evident from Eq. (7.9), the intrinsic sieving coefficient is a function of the permeate flux. The effect of permeate flux on the intrinsic sieving coefficient is shown in Fig. 7.1. S_i decreases as J_v is increased. At very high J_v values, S_i approaches an asymptotic sieving coefficient (S_∞). S_∞ depends on the protein, the membrane and on the physicochemical conditions (i.e., pH, salt concentration, and presence of other proteins).

The intrinsic sieving coefficient of a charged protein with respect to a charged membrane may be expressed as shown below [9]:

$$S_i = S_{i,steric} \exp\left(-\frac{E}{KT}\right) \tag{7.10}$$

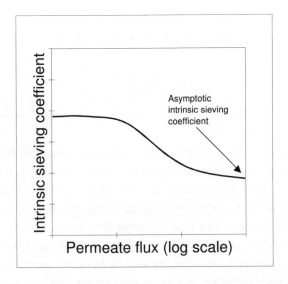

Fig. 7.1. Effect of permeate flux on intrinsic sieving coefficient.

where

$S_{i,steric}$	=	intrinsic sieving coefficient in absence of electrostatic interaction
E	=	activation energy term
K	=	Boltzmann constant
T	=	absolute temperature

The term (E/KT) can be expressed in the form [3]:

$$\left(\frac{E}{KT}\right) = A_a\sigma_p^2 + A_b\sigma_p\sigma_m + A_c\sigma_m^2 \qquad (7.11)$$

where

σ_p	=	surface charge density on protein
σ_m	=	surface charge density on membrane

A_a, A_b and A_c are positive coefficients.

It is often more convenient to use the apparent sieving coefficient (S_a) to present experimental data for protein transmission.

$$S_a = \frac{C_p}{C_b} \qquad (7.12)$$

$$S_a = \frac{S_\infty \exp\left(\frac{S_\infty J_v \delta_m}{D_{\text{eff}}} + \frac{J_v}{k}\right)}{(S_\infty - 1)\left[1 - \exp\left(\frac{S_\infty J_v \delta_m}{D_{\text{eff}}}\right)\right] + S_\infty \exp\left(\frac{S_\infty J_v \delta_m}{D_{\text{eff}}} + \frac{J_v}{k}\right)}$$

$$(7.13)$$

From Eq. (7.13) is evident that for a given protein-membrane system, the apparent sieving coefficient depends on permeate flux, as well as on the mass transfer coefficient. Figure 7.2 shows the effect of permeate flux on the apparent sieving coefficient when the mass transfer coefficient is kept constant. When convective flow is absent (i.e., when $J_v = 0$), the protein is transported through the membrane by diffusion alone. After a sufficient duration, the concentrations on the two sides of the membrane approach equilibrium and, therefore, the theoretical steady state sieving coefficient approaches 1. As the permeate flux is increased, convective transport of protein begins to predominate. The transport of protein molecules by convection is hindered to a greater extent when compared with that of the solvent molecule (i.e., water). The degree of this difference increases with

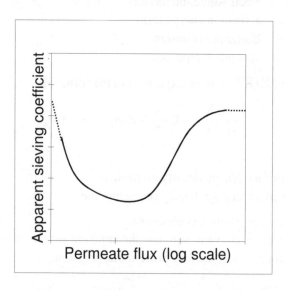

Fig. 7.2. Effect of permeate flux on apparent sieving coefficient.

flow velocity and therefore a decrease in apparent sieving coefficient of the protein is observed with an increase in permeate flux. However, at very high permeate flux values, the wall concentration of the protein increases very significantly. Due to this, the apparent sieving coefficient begins to increase with a further increase in permeate flux and asymptotically approaches a value of 1 at very high J_v values.

The effect of permeate flux on protein transmission has been discussed by van den Berg *et al.* [10]. These workers used 'velocity variation plots' to show the influence of permeate flux, as well as cross flow velocity on the observed rejection coefficient. Using this approach, the mass transfer coefficient for a given operating condition could be determined.

The role of mass transfer coefficient in macrosolute transmission in an ultrafiltration process has been discussed by Gekas and Olund [11]. The effect of mass transfer coefficient on protein transmission in cross-flow ultrafiltration has also been discussed by van den Berg *et al.* [10]. At a particular permeate flux value, the apparent sieving coefficient decreases, as the mass transfer coefficient is increased (see Fig. 7.3). This observed effect could be predicted using Eq. (7.13).

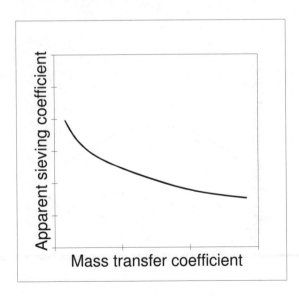

Fig. 7.3. Effect of mass transfer coefficient on apparent sieving coefficient.

pH and salt concentration. Generally speaking, the highest sieving coefficient of a protein is observed when it is uncharged, i.e., at its isoelectric point (pI) (see Fig. 7.4). When a protein is charged (i.e., either has a negative or positive net charge) an electrostatic double layer is formed around the protein molecule by the counter-ions and co-ions present in solution. This so called electrostatic double layer increases the effective size of the protein molecule. However, when a protein carries zero net charge (i.e., when the solution pH corresponds to its pI), the electrical double layer is absent. This is one of the contributory factors towards the highest apparent sieving coefficient of a protein at its pI and lower values on either side on the pH scale.

A membrane could be charged at the operating pH. If the membrane carries a like charge with respect to a protein, the sieving coefficient decreases due to repulsion of the proteins by the membrane. This is referred to as 'intrinsic electrostatic rejection' of the protein. If the membrane has an opposite charge, a 'dynamic membrane' is formed due to the adsorption

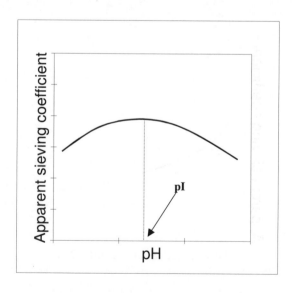

Fig. 7.4. Effect of pH on sieving coefficient.

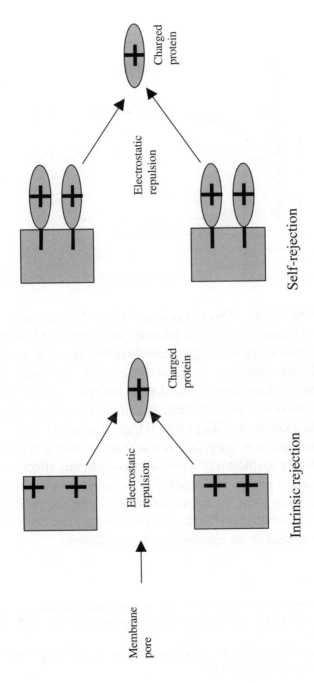

Fig. 7.5. Intrinsic and self-rejection of proteins by membranes.

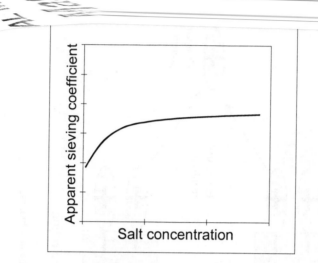

Fig. 7.6. Effect of salt concentration on sieving coefficient of charged protein molecules.

of the protein molecules. This dynamic membrane has the same charge as that on the protein molecules in solution. As a result of this, protein molecules in solution approaching the membrane are repelled and, consequently, the sieving coefficient decreases (see Fig. 7.5). This phenomenon is referred to as 'electrostatic self-rejection' of protein. An uncharged protein does not experience either intrinsic or self-rejection. This is another reason why the transmission of a protein is highest at its pI.

The salt concentration in the feed is also found to have a strong influence on the sieving coefficient of a protein. However, this effect is significant only when the protein is charged. For a charged protein, its sieving coefficient increases when the salt concentration in the feed is increased (see Fig. 7.6). However, with uncharged proteins, the salt concentration has virtually no effect on the apparent sieving coefficient.

References

1. J.D. Ferry, 'Ultrafilter membranes and ultrafiltration' *Chemical Review* **18** (1936) 373.
2. W.M. Deen, 'Hindered transport of large molecules in liquid-filled pores' *AIChE Journal* **33** (1987) 1409.

3. F.G. Smith and W.M. Deen, 'Electrostatic double layer interactions for spherical colloids in cylindrical pores' *Journal of Colloid and Interface Science* **78** (1980) 444.

4. W.R. Bowen and F. Jenner, 'Dynamic ultrafiltration model for charged colloidal dispersions: A Wigner-Seitz cell approach' *Chemical Engineering Science* **50** (1995) 1707.

5. W.R. Bowen and A.O. Sharif, 'The hydrodynamic and electrostatic interactions on the approach and entry of a charged spherical particle to a charged cylindrical pore in a charged planar surface with implications to membrane separation processes' *Proceedings of the Royal Society London A* **452** (1996) 2121.

6. W.R. Bowen and A.O. Sharif, 'Long-range electrostatic attraction between like-charge spheres in a charged pore' *Nature* **392** (1998) 663.

7. O. Kedem and A. Katchalsky, 'Thermodynamic analysis of the permeability of biological membranes to non-electrolytes' *Biochemica Biophysica Acta* 1989 (1958) 413.

8. W.S. Opong and A.L. Zydney, 'Diffusive and convective protein-transport through asymmetric membranes' *AIChE Journal* **37** (1991) 1497.

9. A.L. Zydney and R. van Reis, 'High-performance tangential-flow filtration', W.K. Wang (ed.) *Membrane Separation in Biotechnology*, Marcel Dekker Inc., New York (2001) p. 398.

10. G.B. van den Berg, I.G. Racz and C.A. Smolders, 'Mass transfer coefficients in cross-flow ultrafiltration' *Journal of Membrane Science* **47** (1989) 25.

11. V. Gekas and L. Olund, 'Mass transfer in the membrane concentration polarisation layer under turbulent cross flow II. Application to the characterisation of ultrafiltration membranes' *Journal of Membrane Science* **37** (1988) 145.

Chapter 8

Selectivity of Protein Fractionation in Ultrafiltration

8.1 Selectivity

The efficiency of binary protein fractionation is commonly expressed in terms of the selectivity (ψ) which is defined as [1]:

$$\psi = \frac{S_{a1}}{S_{a2}} \qquad (8.1)$$

Here subscript 1 stands for the preferentially transmitted protein while 2 stands for the preferentially retained protein. The selectivity in ultrafiltration is mainly determined by molecular dimension if the solutes to be separated differ considerably in size. However, for solutes having comparable dimensions (as in protein fractionation), other factors can be made to play significant roles in determining selectivity. For complex protein mixtures (i.e., mixtures made up of more than two proteins), a binary selectivity value cannot be used for defining the efficiency of protein fractionation. A new parameter termed effective selectivity (ψ_E) can be used [2]. This is defined as:

$$\psi_E = \frac{S_{a1}}{S_{a\ Tot}} \qquad (8.2)$$

Where

S_{a1} = apparent sieving coefficient of the protein under consideration

$S_{a\ Tot}$ = apparent sieving coefficient for all other proteins in the mixture

Quite clearly the selectivity would depend on the sieving coefficients of the individual proteins, which in turn would depend on parameters such as permeate flux, system hydrodynamics (i.e., mass transfer coefficient), pH and salt concentration. Attempts have been made to simulate the effects of permeate flux and system hydrodynamics on selectivity in the fractionation of two-protein mixtures [3]. These simulation studies show that for a given cross-flow velocity (i.e., system hydrodynamics), the selectivity depends on the permeate flux, increasing with an increase in permeate flux from a value of 1 to a maximum value (at a permeate flux value J_v^{opt}), and then decreasing to a value of 1 at very high permeate flux values (see Fig. 8.1). The value of J_v^{opt} depends on the cross-flow velocity; the higher the cross-flow velocity, the higher the value of J_v^{opt}. The selectivity is independent of the cross-flow velocity at very low and very high permeate flux values. In the intermediate permeate flux range (at which most ultrafiltration processes are carried out), the selectivity increases with an increase in cross-flow velocity. This cross-flow sensitive range is different for laminar and turbulent flow. In the permeate flux range of 1×10^{-6} m s^{-1} to

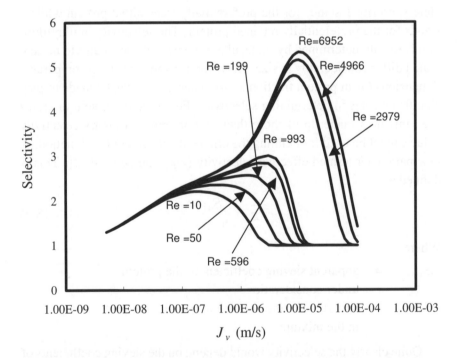

Fig. 8.1. Simulation of the effect of permeate flux on selectivity of two-protein mixture.

1×10^{-5} m s^{-1}, transition from laminar to turbulent flow leads to a very significant increase in selectivity. However, further increase in cross-flow velocity does not increase the selectivity significantly further. For a given system, both permeate flux and cross-flow velocity need to be optimised to obtain high selectivity.

This simulation study was based on the simplifying assumptions that (a) the solutes were non-interacting, (b) the transport parameters for the two solutes (i.e., the free solution diffusivity and diffusivity within the pores) were identical in single protein and binary mixture ultrafiltration, and (c) the solutes had identical sieving properties (i.e., value of S_∞) in single protein and binary mixture ultrafiltration. Whereas these assumptions are reasonable for specific systems, these are not universally valid. Protein–protein interactions can influence the diffusivity of proteins in mixtures. Therefore, while attempting to model transmission behavior in multi-protein ultrafiltration, modified diffusivity values might need to be used. The relative abundance to the preferentially retained protein/s in the concentration polarisation layer may also influence the transmission behaviour of the largely transmitted protein/s. The complex nature of protein–protein interactions has been the main difficulty in modelling and simulation of protein fractionation using ultrafiltration.

8.2   Enhancement of selectivity

The selectivity of separation can be enhanced by means of the following approaches:

a. pH optimisation
b. Feed concentration optimisation
c. Salt concentration optimisation
d. Membrane surface pre-treatment
e. Optimisation of permeate flux and system hydrodynamics

These are discussed in detail in Chapter 12.

References

1. S. Saksena and A.L. Zydney, 'Effect of solution pH and ionic strength on the separation of albumin from immunoglobulins (IgG) by selective filtration' *Biotechnology and Bioengineering* **43** (1994) 960.
2. R. Ghosh and Z.F. Cui, 'Purification of lysozyme using ultrafiltration' *Biotechnology and Bioengineering* **68** (2000) 191.
3. R. Ghosh and Z.F. Cui, 'Simulation study of the fractionation of proteins using ultrafiltration' *Journal of Membrane Science* **180** (2000) 29.

Chapter 9

Protein Concentration

9.1 Introduction

Protein concentration involves the removal of solvent (i.e., water) from protein solutions. A protein concentration process is used to:

a. Increase the protein concentration to facilitate bioseparation steps such as salt and solvent induced precipitation
b. Increase or adjust the concentration of a therapeutic protein in a formulation (e.g. vaccine, monoclonal antibody, therapeutic enzyme)
c. Pre-treat protein solutions for polishing steps, such as crystallisation and freeze drying
d. Increase protein concentration to facilitate detection and analysis

On a small-scale, proteins may be concentrated using different laboratory methods. These include:

a. Vacuum evaporation
b. Dialysis against PEG
c. Dialysis against sucrose
d. Centrifugal ultrafiltration

Vacuum evaporation is very slow and is only feasible with very small sample volumes (i.e., less than 10 ml). Dialysis against PEG is a slow and cumbersome method. It involves the immersion of a sealed dialysis bag filled with protein solution in a concentration solution of PEG. Due to the osmotic pressure difference, water is drawn from inside the dialysis bag to the external solution. Dialysis against sucrose, in addition to being slow and cumbersome, introduces sucrose into the protein samples, which might be undesirable in some cases.

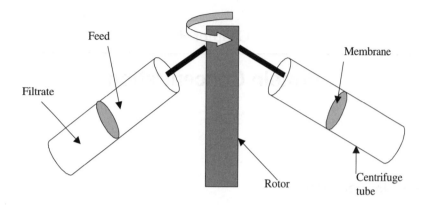

Fig. 9.1. Centrifugal ultrafilter.

Centrifugal ultrafiltration is a small-scale variant of ultrafiltration, where a centrifugal field is used to drive the flow of solvent through a membrane (Fig. 9.1). This technique is specifically intended for processing small volumes of protein solution (i.e., less than 50 ml). The protein solution to be concentrated is loaded in the 'upper' chamber of the modified centrifuge tube and the filtrate or permeate is collected in the 'lower' chamber.

Ultrafiltration is the method of choice for large-scale protein concentration. The protein-water selectivity in ultrafiltration is not a real challenge, due to the significant difference in molecular size. An ideal membrane for protein concentration should totally retain the protein/s while allowing high water permeability through it. The main issue in such operations is achieving satisfactory permeate flux, since this tends to decrease as the feed concentration increases. Also at very high concentrations, proteins tend to form gels, making it difficult to handle them. Another important issue is maintaining the activity of bioactive proteins, i.e., prevention of protein denaturation during processing.

9.2　Solute retention

The type of ultrafiltration membrane to be used for a particular process would depend on the type of protein being concentrated. The membrane is usually selected such that the retention of the protein is 98 percent or

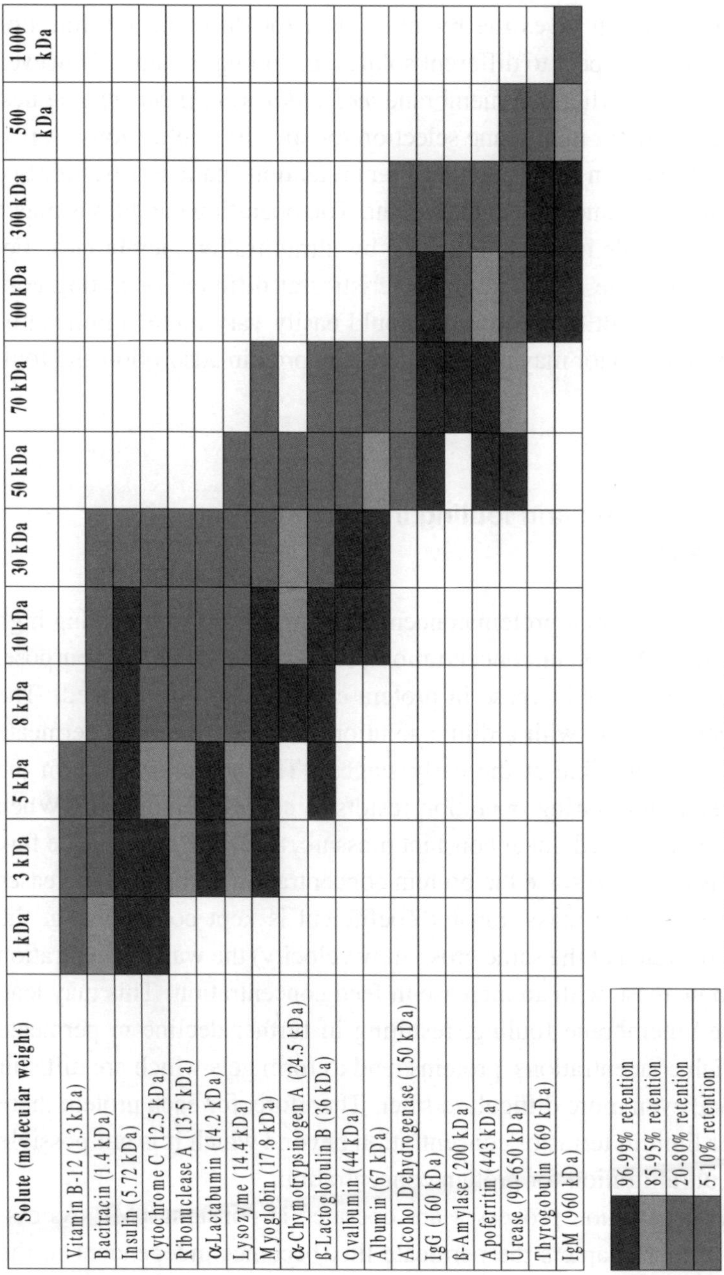

Solute (molecular weight)	1 kDa	3 kDa	5 kDa	8 kDa	10 kDa	30 kDa	50 kDa	70 kDa	100 kDa	300 kDa	500 kDa	1000 kDa
Vitamin B-12 (1.3 kDa)												
Bacitracin (1.4 kDa)												
Insulin (5.72 kDa)												
Cytochrome C (12.5 kDa)												
Ribonuclease A (13.5 kDa)												
α-Lactabumin (14.2 kDa)												
Lysozyme (14.4 kDa)												
Myoglobin (17.8 kDa)												
α-Chymotrypsinogen A (24.5 kDa)												
β-Lactoglobulin B (36 kDa)												
Ovalbumin (44 kDa)												
Albumin (67 kDa)												
Alcohol Dehydrogenase (150 kDa)												
IgG (160 kDa)												
β-Amylase (200 kDa)												
Apoferritin (443 kDa)												
Urease (90-650 kDa)												
Thyrogobulin (669 kDa)												
IgM (960 kDa)												

Legend:
- 96-99% retention
- 85-95% retention
- 20-80% retention
- 5-10% retention

*Data derived using stirred cells, +0.1-0.2% buffered solutions at 3.7 bar (55 psig)

Fig. 9.2. Solute retention characteristics (typical solute rejections*) (Data Courtesy of Pall Corporation).

higher under any operating condition. Figure 9.2 shows a membrane selection guide, which gives the retention data for different ultrafiltration membranes with respect to different solutes, including proteins. However, as was discussed earlier, the membrane *molecular weight cut-off* is at best a rough guideline for membrane selection for specific applications. A particular membrane can show quite different retention characteristics for two solute having the same molecular weight. The operating conditions play a very significant role in solute retention by ultrafiltration membranes. For the same solute, the retention characteristics at different operating conditions (e.g. pH, salt concentration) could easily vary quite significantly. The retention behavior may also be altered by protein adsorption and fouling [1].

9.3　Permeate flux and fouling in concentration processes

The major concern in a protein concentration process is maintaining high permeate flux. A protein concentration process, due to its very purpose, involves a continuous increase in protein concentration in the feed. The process usually starts with a dilute solution and therefore high permeate flux values are possible at the early stages. The non-uniformity in the feed concentration during operation results in several limitations. When the operation is carried out at constant pressure, a decline in permeate flux with time is observed since the protein concentration in the feed increases continuously. If the mass transfer coefficient is kept constant (e.g. the process is operated at the same cross-flow velocity) the wall concentration is likely to increase with an increase in feed concentration. This may lead to increased membrane fouling, resulting in further decline in permeate flux. At high concentrations, proteins tend to form gels which are difficult to pump and even more difficult to filter. Therefore, for each protein there is an upper limit in terms of concentration beyond which it is not possible to rely on ultrafiltration for protein concentration.

Most concentration processes are operated at constant TMP. As discussed in earlier chapters the permeate flux becomes independent of the TMP at very high TMP values. It is therefore sensible to operate below such 'limiting TMP'. At higher TMP the applied pressure in not fully utilised and fouling can be very severe.

When a constant permeate flux is used, the transmembrane pressure increases with time. Eventually the TMP increases to such an extent that further operation is no longer feasible. Constant flux ultrafiltration is rarely used for protein concentration.

Techniques used for permeate flux enhancement (see Chapter 6) have been found to be useful in protein concentration processes.

9.4 Modules

Membrane modules suitable for protein concentration should have the following characteristics:

a. Low hold-up volume on the feed side
b. Ability to handle viscous feed
c. Ability to provide high membrane wall shear rate (or mass transfer co-efficient)

On a laboratory scale, centrifugal ultrafiltration devices are popular. However, as mentioned earlier, these devices can only handle very small liquid volumes. On a slightly larger scale (up to 100 ml), stirred cell ultrafiltration modules are commonly used for protein concentration. These are generally operated dead-ended using constant transmembrane pressure. The transmembrane pressure is generated by pressurising the contents of the feed within the stirred cell with nitrogen or compressed air. Consequently the permeate flux declines with time till virtually nothing passes through the membrane. Such luxury can only be afforded at a laboratory scale. Protein concentration processes on an industrial scale will have to rely on cross-flow filtration.

Flat sheet tangential flow membrane modules are generally preferred for protein concentration. Hollow fibre membranes are used when the viscosity of the final concentrated product is not too high. Tubular membrane devices, which have a very high hold-up volume, are generally not preferred.

9.5   Mode of operation

The most common mode of operation for protein concentration is batch concentration (see Fig. 9.3). A fixed volume of liquid to be concentrated is added to the feed tank and pumped through a cross-flow ultrafiltration membrane module. The filtration is usually operated at constant trans-membrane pressure, which is usually generated by using a pressurising valve on the retentate line. The filtrate (or permeate) is usually discarded, while the retentate is sent back to the feed tank. In a batch concentration process the feed concentration increases with time. If it is feasible to operate at constant permeate flux, the concentration increases as shown in Fig. 9.4 (firm line). However, in a constant pressure operation, the permeate flux would decline with time and therefore the real feed concentration profile would perhaps be more like the dotted line shown in Fig. 9.4. The batch concentration system is simple and efficient. Its efficiency is due to the lowest possible material exposure to the membrane for a given degree of concentration. The main limitations of the batch concentration mode are:

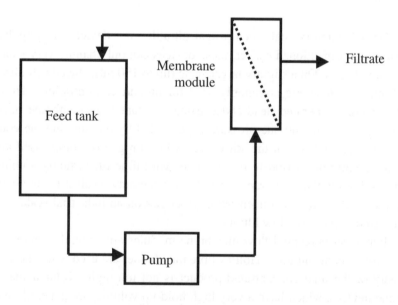

Fig. 9.3. Batch concentration.

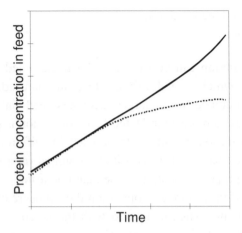

Fig. 9.4. Increase in protein concentration in the batch concentration mode.

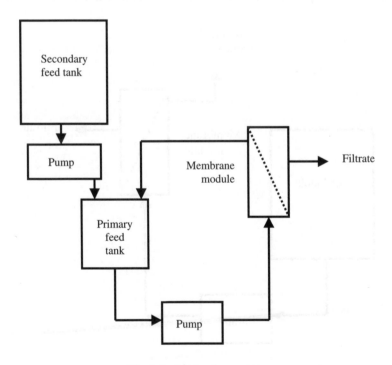

Fig. 9.5. Fed-batch operation.

a. Large dedicated feed tank required
b. Batch operation

The first limitation can be overcome by using a fed-batch operation (see Fig. 9.5) in which a small primary feed tank is used. Into this, feed is continuously pumped from a larger secondary feed tank or, where relevant, directly from a bioreactor. A fed-batch operation is preferred where the same ultrafiltration module is to be used for different concentration operations. A fed-batch set-up is usually mounted on a skid and can be moved to the appropriate location in the manufacturing plant as and when required. In a fed-batch mode of operation the exposure of the membrane to material is greater when compared with the batch mode. Therefore, the permeate flux and efficiency are lower. However, manufacturers often prefer the operational flexibility provided by the fed-batch configuration.

The requirement of a continuous operation is not such an overriding factor in the bioprocess industry. However, when desirable, a continuous

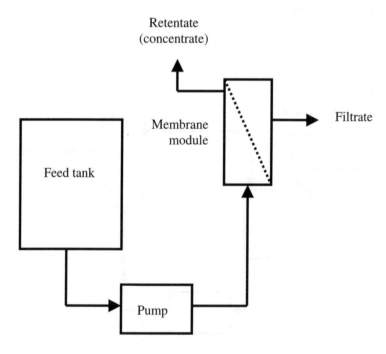

Fig. 9.6. Continuous concentration.

protein concentration operation may be used (see Fig. 9.6). The main limitation of the continuous mode is the low degree of concentration. Consequently, a continuous mode of operation is used only when a protein solution is to be 'slightly' concentrated. When it is necessary to have a continuous process giving a reasonable degree of product concentration, a multi-stage continuous operation may be used (see Fig. 9.7).

Another mode of operation that is frequently employed for large-scale protein concentration is the partial recycle mode (see Fig. 9.8). A certain proportion of the concentrated retentate is directly taken through a recycle line to the feed stream, thus bypassing the feed tank. The main advantage gained from this arrangement is the restriction of viscous concentrate to within a circuit. This avoids complications associated with inefficient mixing of concentrated retentate stream with the relatively dilute liquid within the feed tank.

A variation of the partial recycle mode is sometimes referred to as the feed-and-bleed mode (see Fig. 9.9). The additional advantage of this mode

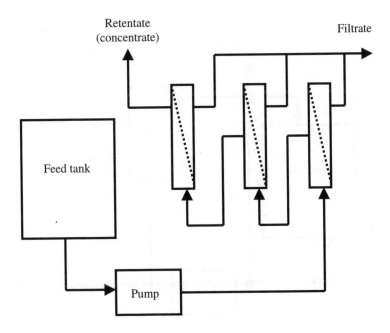

Fig. 9.7. Multi-stage continuous concentration.

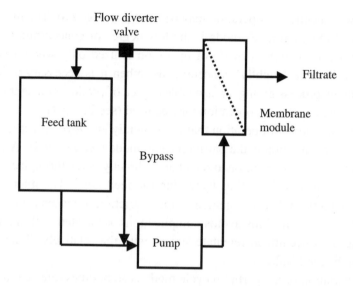

Fig. 9.8. Partial recycle mode for solute concentration.

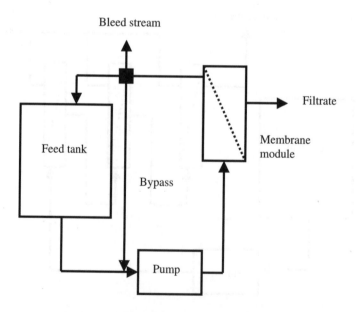

Fig. 9.9. Feed and bleed mode for solute concentration.

of operation is the arrangement for intermittent or continuous 'bleeding' of concentrated retentate stream.

9.6   Applications

Protein concentration is an important and well-established operation in the bioprocess industry. The main areas of application are:

Table 9.1. Use of ultrafiltration for protein concentration.

Protein	Membrane/module	Remarks	Reference
Whey proteins	Cross-flow filtration	Gas sparging used to enhance permeate flux	[2]
Human albumin	30 kDa membrane (cross-flow)	Effect of temperature examined	[3]
Hydrolysed rice-bran proteins	1, 2 and 3 kDa UF membrane (cross-flow)	Effect of chemical treatment examined	[4]
Milk protein	Cross-flow filtration	Preparation of cheese	[5]
Serum albumin	Zirconia and Carbosep membranes (cross-flow)	Ethanol treated albumin	[6]
Serum albumin	Polysulfone and cellulose acetate membranes (cross-flow)	Ethanol treated albumin	[7]
Enzyme	Thin channel and spiral wound membranes		[8]
Cellulases	Polysulfone membrane	Effects of enzyme concentration and TMP examined	[9]
Serine alkaline protease	10 and 30 kDa polysulfone membrane	Effect of TMP, enzyme concentration and cross-flow velocity on permeate flux examined	[10]
Fungal cellulases	10 kDa hollow fibre membrane		[11]

a. Concentration of blood proteins, e.g. albumin, immunoglobulins
b. Processing of milk products, e.g. cheese, milk concentrate
c. Concentration of milk proteins, e.g. casein, α-lactalbumin, β-lactoglobulin
d. Concentration of vegetable proteins, e.g. soy proteins, gluten
e. Concentration of animal enzymes, e.g. amylases, lipases, proteases
f. Concentration of plant enzymes, e.g. papain, bromelain, pectinase
g. Concentration of microbial enzymes, e.g. cellulases, hemicellulases, proteases

Some reports on the use of ultrafiltration for protein concentration are listed in Table 9.1. In a recent paper, Noordman *et al.* [12] discussed protein concentration by ultrafiltration using a Maxwell-Stefan approach. This approach has significant advantages over conventional approaches (e.g. using concentration polarisation model).

References

1. P. Pradanos and A. Hernandez, 'Cross-flow ultrafiltration of proteins through asymmetric polysulfone membranes. 1. Retention curves and pore-size distributions' *Biotechnology and Bioengineering* **47** (1995) 617.
2. A. Martinez-Hermosilla, G.J. Hulbert and W.C. Liao, 'Effect of cottage cheese whey pretreatment and 2-phase crossflow microfiltration/ultrafiltration on permeate flux and composition' *Journal of Food Science* **65** (2000) 334.
3. J. Cabrera-Crespo *et al.*, 'Albumin purification from human placenta' *Biotechnology and Applied Biochemistry* **31** (2000) 101.
4. J.S. Hamada, 'Ultrafiltration of partially hydrolysed rice bran protein to recover value-added products' *Journal of the American Oil Chemists Society* **77** (2000) 779.
5. A.W. Hydamaka, R.A. Wilbey, M.J. Lewis and A.W. Kuo, 'Manufacture of heat and acid coagulated cheese from ultrafiltered milk retentates' *Food Research International* **34** (2001) 197.
6. B.B. Gupta, A. Chaibi and M.Y. Jaffrin, 'Ultrafiltration of albumin ethanol solutions on mineral membranes' *Separation Science and Technology* **30** (1995) 53.
7. M.Y. Jaffrin and J.P. Charrier, 'Optimisation of ultrafiltration and diafiltration processes for albumin production' *Journal of Membrane Science* **97** (1994) 71.

8. C.S. Slater, P.P. Antonecchia, L.S. Mazzella and H.C. Hollein, 'Enzyme concentration and purification — Experimental ultrafiltration studies in thin-channel and spiral wound configurations' *Abstract of Papers of the American Chemical Society* **191** (1986) 150-INDE.

9. M. Pizzichini, C. Fabiani and M. Sperandei, 'Recovery by ultrafiltration of a commercial enzyme for cellulose hydrolysis' *Separation Science and Technology* **26** (1991) 175.

10. S. Takac, S. Elmas, P. Calik and T.H. Ozdamar, 'Separation of the protease enzymes of Bacillus licheniformis from the fermentation medium by cross-flow ultrafiltration' *Journal of Chemical Technology and Biotechnology* **75** (2000) 491.

11. J.C. Roseiro, A.C. Conceicao and M.T. Amaralcollaco, 'Membrane concentration of fungal cellulases' *Bioresource Technology* **43** (1993) 155.

12. T.R. Noordman, T.H. Ketelaar, F. Donkers and J.A. Wesselingh, 'Concentration and desalination of protein solutions by ultrafiltration' *Chemical Engineering Science* **57** (2002) 693.

Chapter 10

Diafiltration of Protein Solutions

10.1 Introduction

Diafiltration is a method by which low molecular weight solutes (e.g. salts, peptide fragments) are removed from protein solutions through ultrafiltration membranes. Diafiltration is also used for buffer exchange, which involves the gradual replacement of the original buffering species by another (see Fig. 10.1). These processes can be carried out by keeping the protein concentration constant or may simultaneously involve concentration or dilution of the protein.

As with protein concentration, the selectivity is not a major issue, on account of the significant difference in solute size. However, the transport of low molecular weight solutes can be influenced by the presence of proteins in the feed.

10.2   Permeate flux and fouling in diafiltration

As with concentration, maintaining high permeate flux is desirable. However, this is less of a challenge in diafiltration, since the protein concentration in the feed does not usually increase with time. In diafiltration it might sometimes be desirable to dilute the feed to enhance the permeate flux.

Diafiltration is usually carried out while keeping the feed volume constant. Therefore, the protein concentration in the feed is also maintained constant. If diafiltration is carried out at constant pressure, the permeate flux may decrease with time due to membrane fouling. Diafiltration can also be carried out at constant permeate flux. This is usually carried out by using a suction pump to draw the permeate through the membrane

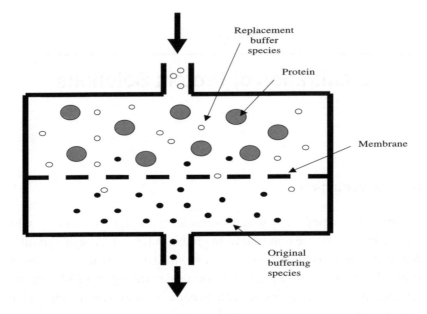

Fig. 10.1. Diafiltration.

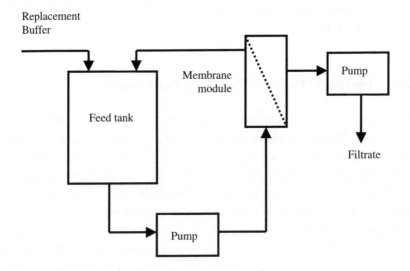

Fig. 10.2. Diafiltration by suction.

(Fig. 10.2). However, the degree of fouling is lower in a constant pressure operation.

10.3 Modules

On a laboratory scale, centrifugal ultrafilters can be used for diafiltration. This is an intermittent operation in which liquid lost through the membrane is replaced by an appropriate buffer. On a slightly larger scale, a stirred cell ultrafilter may be used. This is operated in the dead end mode. The feed is loaded within the stirred cell and the diafiltration buffer is kept in a reservoir connected to the stirred cell. Nitrogen or compressed air is used to pressurise the buffer reservoir and this in turn creates the transmembrane pressure.

Industrial scale diafiltration processes are carried in the cross-flow mode. When compared with concentration processes, a wider variety of membrane modules can be used for diafiltration, i.e., flat sheet tangential flow, hollow fibre, spiral wound and tubular modules may be used.

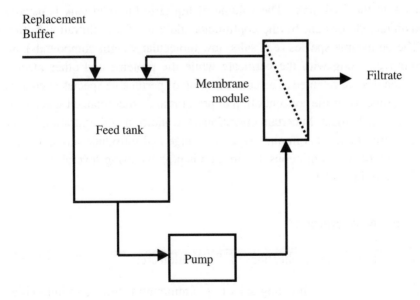

Fig. 10.3. Batch diafiltration.

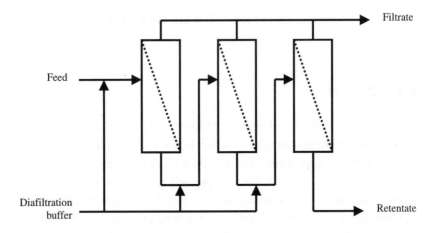

Fig. 10.4. Continuous diafiltration.

10.4   Mode of operation

The most simple and common mode of operation is batch diafiltration (Fig. 10.3). The feed solution is pumped from a feed tank to the ultra-filtration module. The permeate is collected while the retentate is sent back to the feed tank. The volume of liquid in the feed tank is usually maintained constant by the continuous addition of a diafiltration buffer. The permeable species (e.g. salts, low molecular weight compounds) are removed along with the permeate while the proteins and other macro-molecules are retained. The concentration of permeable species decreases with time while the concentration of proteins and other retained species remains unchanged. In certain operations, the addition of diafiltration buffer is intermittent. This implies volume change and introduces complexities to the process. Continuous diafiltration is possible using several stages, as shown in Fig. 10.4.

10.5   Applications

Some of the major applications of diafiltration are:

a. Removal of precipitating salts (e.g. ammonium sulfate, sodium chloride) from protein solutions

b. Removal of precipitating solvents (e.g. ethanol, acetone) from protein solutions
c. Removal of peptide fragments from protein solutions
d. Buffer exchange before and after chromatographic separation
e. Removal of toxic metabolites from blood (i.e., hemodiafiltration)
f. Formulation of proteins in appropriate buffer
g. Removal of inhibitors from enzyme solutions
h. Protein refolding/renaturation

Some of the reports on the application of ultrafiltration for protein diafiltration are listed in Table 10.1. In a recent paper, Noordman *et al.* [10] discussed protein diafiltration by ultrafiltration using a Maxwell-Stefan approach. This approach has significant advantages over conventional approaches (e.g. using concentration polarisation model) and is able to explain the effect of the presence of proteins in solution on the transport of low molecular weight substances through ultrafiltration membranes.

Table 10.1. Application of diafiltration for protein processing.

Protein	Membrane/ module	Remarks	Reference
Whey proteins	Cross-flow	Two-stage batch diafiltration	[1]
Chymotrypsinogen B	Cross-flow	Product formulation	[2]
Lysozyme	Dead-end filtration	Protein refolding by changing urea concentration	[3]
r-Growth differentiation factor 5		Protein refolding by solubilisation of inclusion bodies	[4]
Whey proteins	Multi-stage unit	Water economy studied	[5]
β-lactoglobulin		Change in calcium ion content	[6]
Acid protease		Protein recovery from solid state fermentation	[7]
Recombinant growth hormone		The concentration of SDS adjusted by diafiltration	[8]
Protein hydrolysate		Demineralisation of waste protein hydrolysate	[9]

References

1. D. Barba, F. Beolchini, D. Cifoni and F. Veglio, 'Whey protein concentrate production in a pilot scale two-stage diafiltration process' *Separation Science and Technology* **36** (2001) 587.
2. J. Thommes *et al.*, 'Human chymotrypsinogen B production from Pichia pastoris by integrated development of fermentation and downstream processing. Part 2. Protein recovery' *Biotechnology Progress* **17** (2001) 503.
3. H. Yoshii *et al.*, 'Refolding of denatured and reduced lysozyme with cysteine/cystine red/ox solution in diafiltration' *Journal of Chemical Engineering Japan* **34** (2001) 211.
4. J. Honda, H. Andou, T. Mannen and S. Sugimoto, 'Direct refolding of recombinant human growth differentiation factor 5 for large-scale production process' *Journal of Bioscience and Bioengineering* **89** (2000) 582.
5. D. Barba, F. Beolchini and F. Veglio, 'Minimizing water use in diafiltration of whey protein concentrates' *Separation Science and Technology* **35** (2000) 951.
6. K.R. Kristiansen, J. Otte, R. Ipsen and K.B. Qvist, 'Large-scale preparation of beta-lactoglobulin A and B by ultrafiltration and ion-exchange chromatography' *International Dairy Journal* **8** (1998) 113.
7. H.M. Fernandez-Lahore, E.R. Fraile and O. Cascone, 'Acid protease recovery from a solid-state fermentation system' *Journal of Biotechnology* **62** (1998) 83.
8. H.S. Jeh, C.H. Kim, H.K. Lee and K. Han, 'Recombinant flounder growth hormone from Escherichia coli: Overexpression, efficient recovery, and growth-promoting effect on juvenile flounder by oral administration' *Journal of Biotechnology* **60** (1998) 183.
9. J. Cakl, J. Sir, L. Medvedikova and A. Blaha, 'Demineralization of protein hydrolysates from enzymatic hydrolysis of leather shavings using membrane diafiltration' *Separation Science and Technology* **33** (1998) 1271.
10. T.R. Noordman, T.H. Ketelaar, F. Donkers and J.A. Wesselingh, 'Concentration and desalination of protein solutions by ultrafiltration' *Chemical Engineering Science* **57** (2002) 693.

Chapter 11

Protein Clarification

11.1 Introduction

Protein clarification refers to the removal of particulate matter from protein solutions. The objective of a membrane based protein clarification process is the efficient removal of particulate matter, along with high protein recovery. Protein clarification is encountered in:

a. Sterile filtration of therapeutic protein solutions before dispensing into vials
b. Sterile filtration of therapeutic protein solutions prior to parenteral administration
c. Removal of cell debris after cell disruption
d. Separation of cells from extracellularly secreted proteins
e. Continuous removal of protein products from bioreactors

Microfiltration is perhaps more widely used for clarification processes. However, in several applications (e.g. virus removal), ultrafiltration has obvious advantages over microfiltration. Ultrafiltration is rarely used when microbial cells, cell debris and similar sized particles have to be removed from protein solutions.

11.2 Protein transmission

The requirement for protein transmission in a clarification process is contrary to that in operations such as protein concentration and diafiltration. In a clarification process, high transmission (ideally 100 percent) is desirable. As discussed in earlier chapters, the extent of transmission of a protein through a particular membrane depends on operating and physicochemical parameters. These have to be taken into consideration when selecting a membrane and operating conditions for a particular application.

11.3   Permeate flux and fouling in protein clarification

The low permeate flux of ultrafiltration makes it unattractive when compared with microfiltration for general use in clarification processes. However, microfiltration membranes have been found to be unsatisfactory for virus removal. One of the earliest reports on the use of ultrafiltration for virus removal from protein solutions was by Bechtel *et al.* [1]. The removal of virus from protein solutions by a new type of composite ultrafiltration membrane was reported by DiLeo *et al.* [2]. This work has been widely cited and may be considered to be a key publication in the area of virus removal using ultrafiltration.

Ensuring high permeate flux during a protein clarification process is essential. However, in virus removal processes the emphasis shifts to virus retention by the membrane. Given the greater efficiency of virus removal by ultrafiltration membranes (in comparison to microfiltration membranes), the low permeate flux of ultrafiltration is considered to be a trade-off and therefore tolerated.

In clarification processes there is a lower degree of concentration polarisation (due to transmission of proteins through the membrane). Therefore, the extent of protein adsorption on the external surface of the membrane is likely to be lower than in concentration and diafiltration. However, the proteins enter the pores and this exposes the walls of the pores for adsorption. Generally speaking, the extent of protein induced fouling observed in clarification process is lower than in diafiltration and concentration.

11.4   Particle transmission through UF membrane

The removal of virus particles from protein solutions by ultrafiltration is due to geometric exclusion, i.e., the pore inlet diameter is smaller than the particle. Madaeni [3] has examined the mechanism of virus removal by microfiltration and ultrafiltration membranes. The efficiency of virus removal is expressed in terms of the log removal value or LRV:

$$LRV = \log \left(\frac{N_F}{N_P} \right) \qquad (11.1)$$

where

N_F = Number of virus particles per unit volume of feed
N_P = Number of virus particles per unit volume of permeate

The major concern regarding the use of ultrafiltration membranes for virus removal is the possibility that a few virus particles could pass through. Membranes usually have a pore size distribution and therefore such undesirable virus transmission takes place through abnormally large pores. Surface imperfections and damage could also lead to virus transmission. One approach to solving this problem is to develop membranes with narrow pore size distribution and minimum surface imperfections. Bellara et al. [4] have discussed a technique for pretreating conventional membranes (i.e., with broad pore size distribution) by blocking the abnormally large pores using latex particles. These pre-treated membranes were found to be suitable for virus removal.

11.5 Modules

Membrane modules suitable for protein clarification should have the following characteristics:

a. Low hold-up volume on the feed side
b. Ability to provide high membrane wall shear rate (or mass transfer coefficient)

On a laboratory scale, centrifugal ultrafiltration devices are popular. However, these can only handle very small liquid volumes. On a slightly larger scale (up to 100 ml) stirred cell ultrafiltration modules are commonly used for protein clarification. These are generally operated dead-ended using constant transmembrane pressure. The transmembrane pressure is generated by pressurising the contents of the feed within the stirred cell with nitrogen or compressed air.

On an even larger scale, cartridge type ultrafilters are frequently used. These are also operated in a dead-ended mode. The major problem with dead-ended ultrafiltration is the blocking of pores by particles. Concentration polarisation is also of concern.

Tubular and hollow fibre ultrafiltration modules are preferred for large-scale operations. These are used in the cross-flow mode. Cross-flow minimises the concentration polarisation of proteins. It also prevents the accumulation of particles near the membrane surface by sweeping them away, thus reducing the chance of pore blocking. Another mechanism of removal of particles from close proximity of the membrane to the bulk feed is by inertial lift.

11.6   Mode of operation

Clarification processes are frequently carried out in the dead end mode. This is primarily due to tradition rather than due to any practical reasons.

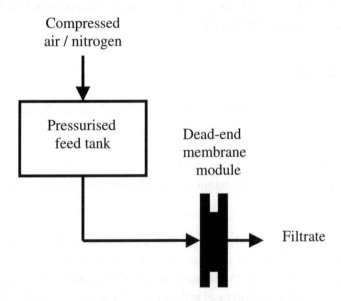

Fig. 11.1. Dead-end clarification.

The only likely advantage of the dead-ended mode of operation is the low hold-up volume on the feed side. Figure 11.1 shows the set-up used for normal dead-ended clarification. If a protein is freely transmitted through the membrane being used, a normal dead-ended mode can give reasonably high recovery of this protein. However, if the protein is partially retained, the recovery with such a set-up would be reduced and there might be extensive fouling. The problem of recovery could be overcome using a dead-ended operation with continuous or intermittent dilution of the feed, as shown in Fig. 11.2. However, the problem of fouling still persists. As mentioned earlier, a major problem with the dead-ended mode is the blocking of pores by particles.

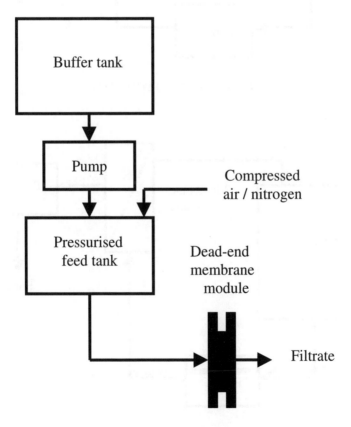

Fig. 11.2. Dead-end clarification with dilution.

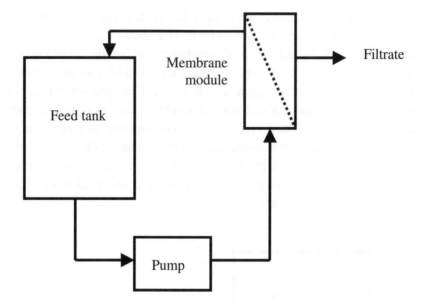

Fig. 11.3. Cross-flow clarification.

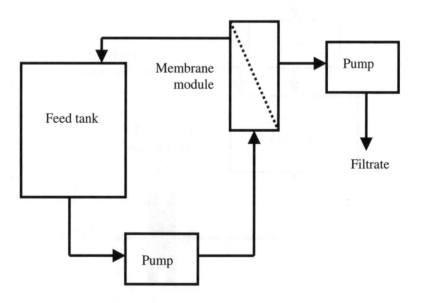

Fig. 11.4. Cross-flow clarification by suction.

Table 11.1. Use of ultrafiltration for virus removal from protein solutions.

Protein	Virus	Membrane/modules	Reference
Protein	30–70 nm virus particles	100 kDa	[2]
Lysozyme, HSA and human IgG	Bacteriophage ΦX-174	100 kDa membrane (flat sheet)	[4]
Factor IX and Factor XI	HIV-1, bovine viral diarrhoea virus	Asahi Planova 15N and 35N (dead ended), Cuprammoniun-regenerated cellulose hollow fibres	[5]
Diaspirin cross-linked human haemoglobin	Bacteriophage ΦX-174, EMC	100 kDa and 500 kDa hollow fibre, Asahi Planova 75N, 35N and 15N (dead-ended), Pall Ultipor VF-UDV50 (dead-ended)	[6]
Monoclonal antibody	Murine leukaemia virus	Pall Ultipor VF-UDV50 (dead-ended), Millipore Viresolve 70 kDa and 180 kDa (tangential flow), Asahi Planova 75N, 35N and 15N (dead-ended)	[7]
BSA, Monoclonal IgG	Herpes simplex, poliovirus, vaccinia virus	Pall Ultipor VF-UDV50 (dead-ended)	[8]
Factor IX	Various viruses	Asahi Planova 35N and 15N (dead-ended)	[9]
Various protein containing media	Phage T1, Phage PP7, poliovirus	6 kDa, 13 kDa and 50 kDa hollow fibres	[10]
Diaspirin cross-linked haemoglobin	Bacteriophage ΦX-174, EMC, HIV, pseodorabies	100 kDa and 500 kDa hollow fibre (A/G technology)	[11]
Monoclonal antibody	Murine retrovirus		[12]

Most of the problems associated with the dead-ended mode of operation can be overcome using the cross-flow mode. Figure 11.3 shows a typical set-up used for a cross-flow clarification process. The protein is the transmitted species while the particles are the retained species. The transmembrane pressure is usually generated by throttling the retentate line. A cross-flow clarification process could also be operated using a suction pump on the permeate line as shown in Fig. 11.4.

11.7 Applications

The use of ultrafiltration for protein clarification is primarily in the area of virus removal. However, with the increasing risk of the contamination of therapeutic products by prions, the use of ultrafiltration for removal of these substances from protein solutions might be an option worth considering. Some of the reports on the application of ultrafiltration for virus removal from protein solutions are listed in Table 11.1.

References

1. M.K. Bechtel, A. Bagdasarian, W.P. Olson and T.N. Estep, 'Virus removal or inactivation in hemoglobin-solutions by ultrafiltration or detergent solvent treatment' *Biomaterials, Artificial Cells and Artificial Organs* **16** (1988) 123.
2. A.J. DiLeo, A.E. Allegrezza Jr and S.E. Builder, 'High-resolution removal of virus from protein solutions using a membrane of unique structure' *Bio/Technology* **10** (1992) 182.
3. S.S. Madaeni, 'Mechanism of virus removal using membranes' *Filtration and Separation* **34** (1997) 61.
4. S.R. Bellara, Z.F. Cui, S.L. MacDonald and D.S. Pepper, 'Virus removal from bioproducts using ultrafiltration membranes modified with latex particle pre-treatment' *Bioseparation* **7** (1998) 79.
5. M. Burnouf-Radosevich, P. Appourchaux, J.J. Huart and T. Burnouf, 'Nanofiltration a new specific virus elimination method applied to high purity factor IX and factor XI concentrates' *Vox Sang* **67** (1994) 132.
6. K.F. Ogle and M.R. Azari, 'Virus removal by ultrafiltration', W.K. Wang (ed.) *Membrane Separations in Biotechnology* 2nd ed., Marcel Dekker Inc., New York (2001) p. 299.
7. P.Y. Huang and J. Peterson, 'Scaleup and virus clearance studies on virus filtration in monoclonal antibody manufacture', W.K. Wang (ed.) *Membrane Separations in Biotechnology* 2nd ed., Marcel Dekker Inc., New York (2001) p. 327.

8. P. Roberts, 'Efficient removal of viruses by a novel polyvinylidene fluoride membrane filter' *Journal of Virological Methods* **65** (1997) 27.
9. A. Johnston *et al.*, 'Inactivation and clearance of viruses during the manufacture of high purity Factor IX' *Biologicals* **28** (2000) 129.
10. K.H. Oshima, T.T. Evansstrickfaden, A.K. Highsmith and E.W. Ades, 'The removal of phages T1 and PP7, and poliovirus from fluids with hollow-fiber ultrafilters with molecular-weight cutoffs of 50000, 13000 and 6000' *Canadian Journal of Microbiology* **41** (1995) 316.
11. M. Azari *et al.*, 'Evaluation and validation of virus removal by ultrafiltration during the production of diaspirin crosslinked haemoglobin' *Biologicals* **28** (2000) 81.
12. W. Berthold, J. Walter and W. Werz, 'Experimental approaches to guarantee minimal risk of potential virus in purified monoclonal antibodies' *Cytotechnology* **9** (1992) 189.

Chapter 12

Protein Fractionation

12.1 Introduction

Protein fractionation (i.e., protein–protein separation) using ultrafiltration is strongly influenced by operating and physicochemical parameters and hence such processes need be very precisely 'fine-tuned' to achieve satisfactory level of separation, e.g. [1– 47]. The fine-tuning exercise includes optimisation of pH, salt concentration, permeate flux and system hydrodynamics. Protein fractionation is much more demanding than processes such as protein concentration and desalting. Significant research work in this area has only been done in recent years and a substantial amount of work still needs to be done in order to 'perfect the technology'.

12.2 Challenges facing ultrafiltration based protein fractionation

Achieving good selectivity is a challenge, mainly due to the following reasons:

12.2.1 *Broad pore size distribution*

The broad pore size distribution observed with most commercial membranes means that the molecular weight cut off (MWCO) value indicated by the manufacturer can only be taken as an approximate guideline in selecting a membrane for a particular application.

12.2.2 *Concentration polarisation*

Concentration polarisation, which refers to the accumulation of protein molecules near the membrane surface, can change the apparent sieving coefficient of the proteins. Concentration polarisation also results in the lowering of permeate flux and may promote adsorption and fouling. The selectivity generally decreases as the extent of concentration polarisation increases. In multi-protein ultrafiltration, the presence of the largely retained proteins in the concentration polarisation layer could have significant effect on the transmission of the preferentially transmitted proteins.

12.2.3 *Fouling/adsorption*

Adsorption of a protein on a membrane and subsequent fouling may alter its sieving properties with respect to the protein. Fouling and adsorption can decrease both permeate flux and protein transmission. In certain cases, however, selective adsorption of specific proteins on ultrafiltration membranes has been found to be beneficial for protein fractionation [14, 46].

12.2.4 *Protein–protein interactions*

Proteins are multipolar macromolecules with carboxyl and amino groups. A protein in solution shows a net electrostatic charge depending on the solution pH. Protein molecules are positively charged when the solution pH is below their pI and negatively charged when the solution pH is above their pI. Oppositely charged molecules may interact electrostatically, thus reducing protein transmission through a membrane [5]. Other forms of interactions, e.g. van der Walls forces and hydrophobic interactions are also known to exist. These may result in protein aggregation and make it difficult to fractionate proteins by ultrafiltration.

12.2.5 *Effects of physicochemical and process parameters on transmission*

The transmission of a protein through a membrane depends on physicochemical parameters such as pH and salt concentration and process parameters such as permeate flux and system hydrodynamics. The effects of these parameters on transmission of different proteins differ, largely depending on the properties of each individual protein and the membrane being used. This implies that selectivity will be strongly influenced by physicochemical and operating parameters and these may need to be optimised. On the other hand, this gives us several handles for optimising a protein fractionation process.

12.3   Selectivity enhancement

Possible ways by which selectivity of protein fractionation can be enhanced are:

12.3.1   *Proper membrane selection and membrane surface modification*

The heart of a membrane separation process is the membrane itself and therefore selection of UF membranes with correct surface and pore properties is important for the success of a protein fractionation process. Research work done in this area clearly illustrates this point. Higuchi *et al.* [4] selected a surface modified polysulfone membrane having a MWCO of 200 kDa to fractionate IgG (mol. wt. 150000, p$I \sim$ 7) and BSA (mol. wt. 67000, pI 4.7). They found that using this membrane fractionation was feasible at pH 8.0. At this pH BSA was nearly totally rejected while IgG was transmitted through the membrane. This type of phenomenon is termed 'reversed selectivity'; a situation where the smaller protein is rejected by a membrane while the larger protein passes through. Miyama *et al.* [19] used positively charged membranes prepared from polyacrylonitrile grafted poly-(N,N-dimethylaminoethyl methacrylate) for the separation of BSA and IgG and found these membranes to be suitable for fractionation. The efficiency of fractionation was also shown to be dependent on physicochemical parameters.

In certain situations, combination of more than one membrane might be useful for protein fractionation. Perepechkina and Perepechkin [12] demonstrated the feasibility of fractionation of leukocyte extract components using a 15 kDa MWCO and a 5 kDa MWCO membrane pair in a two-step ultrafiltration process. Ghosh and Cui [47] have reported a 50 kDa and

25 kDa MWCO polysulfone membrane combination based two-step ultra-filtration process for high-resolution purification of lysozyme from chicken egg white. In this process, the first step (i.e., using the 50 kDa MWCO membrane) gave a partially purified lysozyme, which was further purified in the second step to yield lysozyme of pharmaceutical grade purity. This combination gave high productivity, along with high purity.

Surface modification of membranes can significantly affect protein fractionation. Legallais *et al.* [21] showed that the adsorption of lipoproteins and albumin on cellulose diacetate membranes altered the selectivity in a plasma fractionation process. Millesime *et al.* [20] reported the fractionation of BSA and lysozyme (mol. wt. 14100, pI 11.0) using unmodified and chemically modified inorganic ultrafiltration membranes. With the unmodified membrane, the selectivity was found to depend on the ionic strength, while with the modified membrane, the selectivity was

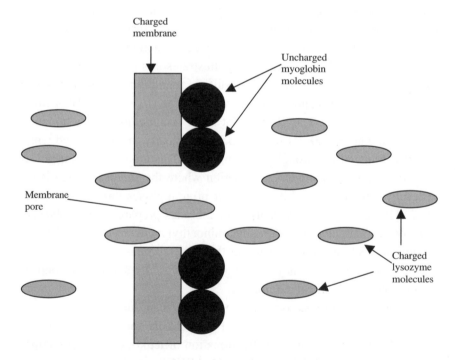

Fig. 12.1. Membrane surface pretreatment by myoglobin to increase lysozyme transmission.

found to be less sensitive to the ionic strength. Thus, with the modified membrane, there was greater operational flexibility. Ghosh and Cui [14] have reported a membrane modification method, which gave enhanced fractionation of BSA and lysozyme using a 50 kDa MWCO polysulfone membrane. The membrane was pre-treated by passive adsorption of another protein myoglobin. Figure 12.1 shows how the bound uncharged myoglobin molecules minimised self-rejection of lysozyme molecules. Enhanced transmission of lysozyme and hence enhanced selectivity was observed with the pre-treated membrane (data shown in Table 12.1). More recently, Ghosh and Cui [46] have reported the enhancement of lysozyme purification from chicken egg white, using similarly surface pre-treated membranes (data shown in Table 12.2).

12.3.2   *pH and salt concentration optimisation*

As mentioned earlier, pH and salt concentration play critical roles in protein fractionation. Both, especially pH, affect the protein–protein and protein-membrane interactions. Thus, these physicochemical parameters need to be optimised. Ohno *et al.* [2] reported excellent separation of IgG and BSA using a 100 kDa MWCO polysulfone membrane by adjusting the

Table 12.1. Enhanced fractionation of BSA and lysozyme using pre-treated membrane (after Ghosh and Cui [14]).

Parameter	Native (untreated) membrane	Pre-treated membrane
Transmembrane pressure (kPa)	20	20
Membrane	50 kDa MWCO polysulfone	Same membrane treated with myoglobin
Feed	2 kg m^{-3} BSA $+ 2 \text{ kg m}^{-3}$ lysozyme, pH 7.0	2 kg m^{-3} BSA $+ 2 \text{ kg m}^{-3}$ lysozyme, pH 7.0
Stirring speed (rpm)	1400	1400
Permeate flux $(\text{kg m}^{-2} \text{ s}^{-1})$	0.00254	0.00281
S_a for BSA	0.0093	0.0067
S_a for lysozyme	0.6776	0.9041
Selectivity	75.82	139.89

Table 12.2. Purification of lysozyme using pre-treated membrane (after Ghosh and Cui [46]).

TMP (kPa)	Permeate flux (m s^{-1})	S_a lysozyme*	S_a ovalbumin*	Percent purity of lysozyme*	Purification factor*
20	2.31×10^{-6}	1.005	0.250	18.1	5.3
40	4.00×10^{-6}	0.952	0.175	24.0	7.1
60	4.89×10^{-6}	1.014	0.038	57.7	16.9
80	5.29×10^{-6}	1.009	0.015	78.8	23.2
100	5.54×10^{-6}	1.004	0.007	89.2	26.2
120	5.59×10^{-6}	0.998	0.004	96.6	28.4
140	5.72×10^{-6}	0.992	0.004	95.8	28.2
160	5.71×10^{-6}	0.993	0.005	94.2	27.7

*Based on FPLC® results.
Membrane: 50 kDa MWCO polysulfone pre-treated by passive adsorption of myoglobin
Feed: 10 kg m^{-3} chicken egg white
Buffer: 20 mM sodium phosphate +100 mM NaCl, pH 7.0
Stirring speed: 1400 rpm
Pure water flux: 20.3–20.9 $\times 10^{-6}$ m s^{-1} at 100 kPa TMP
S_a of lysozyme was about 26 percernt higher than with untreated membrane.

pH between 4 and 5 and the salt concentration below 0.2 M. These physicochemical conditions were found to be critical for fractionation. Zhang and Spencer [22] have also stressed the need for pH and salt concentration 'fine-tuning' in the fractionation of bovine gamma globulin and bovine albumin by filtration with porous stainless steel supported titanium dioxide formed-in-place membranes. Saksena and Zydney [6] reported similar effects while investigating the fractionation of IgG and BSA using 100 kDa and 300 kDa MWCO polyethersulfone membranes. They explained these effects in terms of the electrostatic contributions of the protein molecules to both bulk (external) and membrane transport.

Iritani *et al.* [7] have shown that solution pH, ionic strength and concentration of BSA in the feed influenced the fractionation of lysozyme and BSA, using a 30 kDa MWCO polysulfone membrane. Ghosh and Cui [14] have also reported the effects of pH on fractionation of BSA and lysozyme using a 50 kDa MWCO polysulfone membrane (data shown in Table 12.3). At pH 5.2, selectivity was 3.3, while at pH 8.8 a selectivity of 220 was observed when all other parameters were kept constant. This clearly shows that pH is a powerful handle for optimising protein fractionation processes.

Table 12.3. Effect of pH on fractionation of BSA and lysozyme (after Ghosh and Cui [14]).

pH	S_a BSA	S_a lysozyme	Selectivity
5.2	0.2500	0.8540	3.3
6.0	0.0400	0.9610	24.0
7.0	0.0059	0.9890	167.6
8.0	0.0050	0.9810	196.2
8.8	0.0045	0.9898	220.0

Membrane: PS, 50 kDa MWCO
Feed: 0.5 kg m^{-3} BSA + 0.5 kg m^{-3}
   lysozyme
Buffer: 20 mM phosphate
TMP: 20 kPa
Stirrer speed: 1400 rpm

In a recent paper, Ghosh and Cui [47] showed that pH optimisation can have a profound influence on the purification of lysozyme from egg white (see Table 12.4).

Nakatsuka *et al.* [23] studied the fractionation of myoglobin (mol. wt. 17000, pI 7.0) and BSA using non-sorptive regenerated cellulose membranes (30 kDa MWCO), and sorptive polysulfone membrane (30 kDa MWCO) and found the selectivity to be relatively independent of ionic strength but strongly dependent on pH. Nakao *et al.* [3], while investigating the separation of myoglobin and cytochrome C (mol. wt. 12300, pI 10.65) using charged membranes, found that the sieving coefficients of proteins were strong functions of pH. The sieving coefficient of

Table 12.4. Effect of pH on purification of lysozyme from chicken egg white (after Ghosh and Cui [47]).

pH	Purification factor	Percent purity of lysozyme
4.6	8.8	30.1
7.0	18.6	69.0
9.0	18.1	66.7
11.0	26.6	88.7

Membrane: PS, 50 kDa MWCO
Feed: 10 kg m^{-3} chicken egg white in buffer
Buffer: 20 mM phosphate, pH adjusted with NaOH
TMP: 80 kPa

each protein was found to be highest at its respective isoelectric pH value. Sannier *et al.* [24] have reported similar findings, based on their study on the effects of pH and ionic strength on fractionation of fish myoglobin and haemoglobin. A scheme for optimising selectivity of binary protein fractionation based on pH effect is shown in Fig. 12.2.

Sudareva *et al.* [5] examined the effects of solution pH on separation of human serum albumin (mol. wt. 65000, pI 4.9) from cytochrome C, and ribonuclease A (mol. wt. 13700, pI 9.45) from haemoglobin (mol. wt. 66080). The selectivity obtained with binary mixtures was considerably lower than those predicted from single protein ultrafiltration experiments. This was ascribed to the formation of complexes due to the electrostatic interaction between unlike charged protein molecules in bulk solution. These effects could be minimised by changing the pH and salt concentration. van Eijndhoven *et al.* [8] studied the role of electrostatic interactions on the fractionation of albumin and haemoglobin using a 100 kDa MWCO omega (polyethersulfone) membrane and reported that higher selectivity was obtained at lower salt concentration. The fractionation process was based on the rejection of a specific protein due to protein-membrane electrostatic

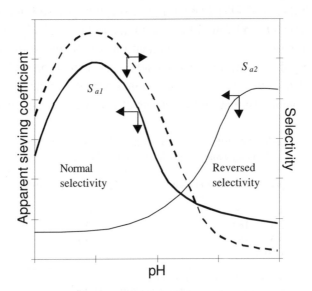

Fig. 12.2. An idealised representation of the possibility of using pH effect for protein fractionation using ultrafiltration.

interactions. These interactions were found to diminish at high salt concentrations.

Mehra *et al.* [25] have discussed the effects of pH, ionic strength and calcium chelation on permeate flux and protein transmission while filtering whey proteins through a 100 kDa MWCO membrane. Fractionation of low molecular weight proteins (e.g. beta-lactoglobulin and alpha-lactalbumin) from high molecular weight proteins (e.g. BSA, lactoferrin and immunoglobulins) was best achieved at pH 8.0. Kawasaki *et al.* [26] discussed the fractionation of kappa-casein glycomacropeptide from whey protein concentrate utilising the pH dependent change in apparent molecular weight of the protein. Nau *et al.* [27] found that ionic interactions between peptides and membrane affected the fractionation of tryptic hydrolysates of beta-casein using 10 kDa MWCO Carbosep M5 membrane. Peptides having same charge as that of the membrane were found to have lower transmission than that expected from the size exclusion model, while peptides with opposite sign had higher transmission than expected. Ehsani *et al.* [10] discussed the effects of pH on the separation of beta-xylanase from a mixture of xylanases and cellulases by ultrafiltration with polysulfone membranes. High selectivity of beta-xylanase purification with PM-30 membrane was observed at pH 7.0. In another paper, Ehsani *et al.* [11] studied the pH dependent fractionation of ovalbumin from egg proteins using 50 kDa MWCO polysulfone membrane. At lower pH and salt concentration, contaminating proteins were rejected due to electrostatic exclusion from the pores, while ovalbumin was transmitted through the membrane.

12.3.3 *Concentration polarisation control*

The accumulation of the preferentially rejected proteins near the membrane surface will generally reduce the transmission of the preferentially transmitted proteins. Severe concentration polarisation will promote protein–protein interaction and protein adsorption on the membrane surface and may even lead to gel layer formation. Control of concentration polarisation is important in maintaining the actual selectivity of the membrane.

Slater *et al.* [28] investigated the fractionation of bovine alkaline phosphatase and BSA with a 100 kDa MWCO regenerated cellulose membrane using a stirred cell ultrafiltration unit. Fractionation was found to be dependent on the stirring speed; separation improved with increase in stirring speed, i.e., improvement in system hydrodynamics. Balakrishnan and Agarwal [48] investigated the role of system hydrodynamics and solution environment on transmission of proteins through a 100 kDa MWCO hydrophilic polyacrylonitrile membrane, using a vortex flow device. The hydrodynamic parameters investigated were permeate flux, axial velocity and rotation speed. Lower protein transmission was observed at higher rotation speeds. The effect of flux on transmission was also demonstrated and explained in terms of a combined concentration polarisation-irreversible thermodynamic model. In a subsequent paper [9], the effects of these process parameters on fractionation of lysozyme from ovalbumin, and lysozyme from myoglobin were investigated. The trends observed in the single protein ultrafiltration experiments, i.e., [48] correlated well with fractionation experiments, i.e., [9]. Krishnan *et al.* [29] reported the use of 100 kDa and 300 kDa MWCO membranes for separation of monoclonal IgM antibodies obtained using hybridoma cell culture technology. Tangential flow velocity and transmembrane pressure were found to affect IgM transmission. Najarian and Bellhouse [30] studied the effect of pressure pulsation on the fractionation of bovine plasma proteins using a flat sheet ultrafilter equipped with ladder like flow deflectors. Application of transmembrane pressure pulsation reduced concentration polarisation and enhanced fractionation.

Li *et al.* [16] reported the effects of gas sparging (a novel concentration polarisation controlling method) on the separation of HSA and IgG using 100 kDa MWCO PES flat sheet membrane. Gas sparging, which involves the injection of gas bubbles into the feed stream (see Fig. 12.3), was found to enhance the permeate flux as well as selectivity of protein fractionation. The beneficial effects of gas sparging on fractionation of HSA-IgG mixture using a tubular membrane module (PVDF, 100 kDa MWCO membrane) was also reported by Li *et al.* [13]. Ghosh *et al.* [15] explained the positive effects of gas sparging on fractionation of BSA and lysozyme using a mass transfer model. The selectivity enhancement data is shown in Table 12.5.

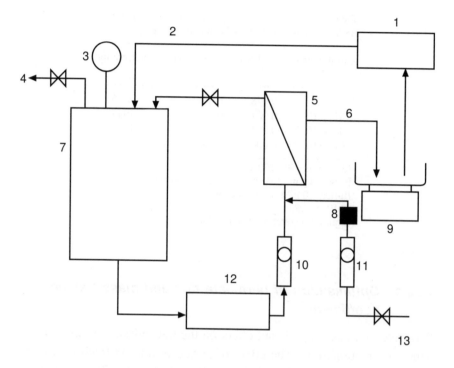

Fig. 12.3. Gas-sparged (two-phase flow) ultrafiltration.

1. Peristaltic pump (for permeate recirculation)
2. Permeate return
3. Pressure gauge
4. Vent line
5. Membrane module
6. Permeate
7. Feed reservior
8. Non-return valve
9. Electronic balance
10. Liquid rotameter
11. Gas rotameter
12. Gear pump (feed pump)
13. Compressed air

Table 12.5. Effect of gas sparging on fractionation of
BSA and lysozyme (after Ghosh *et al.* [15]).

Gas flow rate $m^3 \, s^{-1}$	S_a BSA	S_a lysozyme	Selectivity
0	0.078	0.856	10.9
1.67×10^{-6}	0.004	0.776	194.0
3.33×10^{-6}	0.004	0.785	196.3
6.67×10^{-6}	0.004	0.739	184.8

Membrane: PS, 100 kDa MWCO
Feed: 2 kg m^{-3} BSA + 2 kg m^{-3} lysozyme
Buffer: 20 mM phosphate, pH 7.0
TMP: 50 kPa
Liquid flow rate: 8.33×10^{-6} m^3 s^{-1}

12.3.4   *Optimisation of permeate flux and mass transfer coefficient*

Permeate flux has a significant effect on the transmission of proteins and hence on fractionation. The effect of permeate flux on fractionation has been discussed by several workers, e.g. [6, 8, 9, 17, 18, 47]. If system hydrodynamics is kept unchanged, the selectivity is generally low at very low and at very high permeate flux values. Maximum selectivity is observed at an intermediate permeate flux value $(J_{v \, opt})$ [49]. Figure 12.4 shows the effect of permeate flux on purification of lysozyme from chicken egg white. Figure 12.5 show how the permeate flux of an ultrafiltration based protein fractionation process can be optimised for maximum selectivity. In this optimisation study, the mass transfer coefficient was kept constant.

Mass transfer coefficient (which depends on system hydrodynamics) also has a significant effect on protein transmission and hence on selectivity [13, 15, 16, 17, 49]. Figure 12.6 shows how the wall shear rate (on which the mass transfer coefficient depends) of an ultrafiltration process affects selectivity. The transmission of the preferentially retained protein is more affected than the transmission of the preferentially transmitted protein. Thus, the better the system hydrodynamic, the higher the selectivity. In order to maximise selectivity, both the permeate flux and system hydrodynamics need to be 'fine-tuned'.

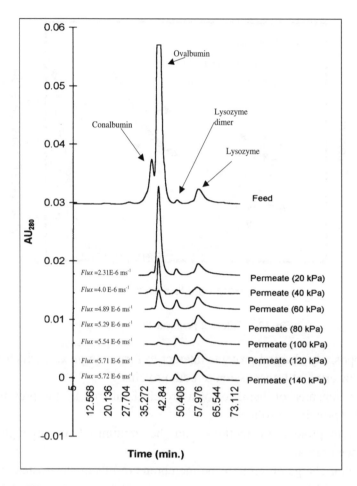

Fig. 12.4. Effect of permeate flux on the purification of lysozyme. The FPLC chromatograms show that as the permeate flux was increased, the purity of lysozyme in the permeate increased (after Ghosh and Cui [46]).

12.4 Thumb rules for protein fractionation

Based on research findings, the following generalised observations can be made:

a. Transmission of a protein is highest at its isoelectric point
b. Effect of pH on protein transmission is negligible at high salt concentrations

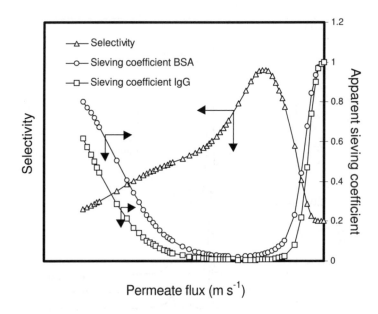

Fig. 12.5. Effect of permeate flux on selectivity.

c. Oppositely charged proteins interact to form complexes, which result in lowering of transmission of both proteins
d. The transmission behavior of a protein is altered by the presence of other proteins in solution
e. Protein–protein interactions can be minimised using high salt concentration
f. Similar charge on protein and membrane results in electrostatic repulsion and hence lower transmission
g. By manipulating the pH and ionic strength, it is possible to obtain 'reversed selectivity', a situation where there is higher transmission of the *larger* protein
h. Protein transmission and hence selectivity depends on permeate flux and mass transfer coefficient (and hence on system hydrodynamics). The permeate flux and system hydrodynamics need to be 'fine-tuned' for optimum selectivity
i. Adsorption of proteins on the membrane surface and within the pores can dramatically alter the flux and transmission properties of a membrane. For instance, adsorption of charged protein molecules on the

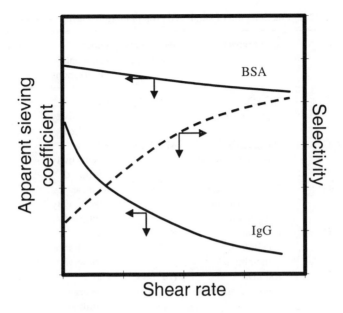

Fig. 12.6. Effect of shear rate on selectivity.

membrane surface leads to 'self rejection' of that protein due to electro-static repulsion of the proteins in solution by membrane bound proteins

12.5 Application areas

Ultrafiltration has been used for a wide variety of protein separations. A large bulk of the literature reviewed deals with the separation of simu-lated binary protein mixtures. These studies are invaluable for develop-ing the 'science of protein fractionation using ultrafiltration'. They also demonstrate the feasibility of using ultrafiltration for protein fractiona-tion to potential users of the technology. Regrettably there are relatively fewer papers dealing with the use of ultrafiltration for fractionation of complex protein mixtures. In Table 12.6, an attempt has been made to summarise the reports on the use of ultrafiltration for the fractionation of simulated protein mixtures. Simulated mixtures that have been fraction-ated are based on well-characterised and easily available proteins, such as BSA, lysozyme, myoglobin, ovalbumin, conalbumin, cytochrome C

Table 12.6. Fractionation of simulated protein mixtures using ultrafiltration.

Proteins	Membrane(s)	Parameters examined	Other comments/ observations	Reference
BSA/IgG	200 kDa polysulfone	M P	Reverse selectivity utilised	[4]
BSA/IgG	100 and 300 kDa polyethersulfone	M P J S C	Reverse selectivity utilised	[6]
BSA/IgG	Titanium dioxide formed-in-place membrane	H P S C	Reverse selectivity utilised	[22]
BSA/IgG	100 kDa cellulosic membrane	P S C	Steric hindrance observed	[36]
HSA/IgG	100 kDa polyethersulfone	H P J S C	Gas sparging utilised	[16]
HSA/IgG	100 kDa PVDF	H P J S C	Gas sparging utilised	[13]
BSA/IgG/ ferritin	7.3 nm pore size glass membrane		Effect of pore size distribution examined	[40]
BSA/ Myoglobin	30 kDa polysulfone and cellulosic membranes	M P J S C		[23]
BSA/ bovine alkaline phosphatase	100 kDa regenerated cellulose	H		[28]
BSA/ lysozyme	30 kDa Amicon PM 30	P S		[1]
BSA/ lysozyme	150 kDa inorganic membrane	M P S		[20]
BSA/ lysozyme	30 kDa polysulfone	P S C		[7]
BSA/ lysozyme	50 kDa polysulfone/ surface modified membrane	P M C	Surface modification using protein	[14]

Table 12.6. *Continued.*

Proteins	Membrane(s)	Parameters examined	Other comments/ observations	Reference
BSA/ lysozyme	100 kDa polysulfone	H J C	Gas sparging utilised	[15]
BSA/ hemoglobin	100 kDa polyacrylonitrile, polysulfone, regenerated cellulose	P H J	Vortex flow device used	[38]
BSA/ hemoglobin	100 kDa polyethersulfone	P J S C		[8]
Hemoglobin/ ribonuclease A	Ripor membranes (copolymer)	P S		[5]
Cytochrome C /HSA	Ripor membranes (copolymer)	P S		[5]
Myoglobin/ cytochrome C	500 kDa polysulfone	M H P J		[3]
Lysozyme/ myoglobin	25 kDa polysulfone	P J	Carrier phase UF utilised	[18]
Lysozyme/ myoglobin	100 kDa polyacrylonitrile	H P J S C	Vortex flow UF device used	[9]
Lysozyme/ ovalbumin	100 kDa polyacrylonitrile	H P J S C	Vortex flow UF device used	[9]
BSA/ myoglobin	50 kDa polysulfone	P M J S C		[37]
Ovalbumin/ myoglobin	50 kDa polysulfone	P M J S C		[37]
Lactoferrin/ BSA	Cellulosic YM- 100 membrane	P M J S C		[37]

Parameters:
M = Membrane surface modification, H = System hydrodynamics,
P = pH, J = Permeate flux, S = Salt concentration,
C = Protein concentration

and chymotrypsinogen. These proteins, in addition to being well characterised, are available in very pure forms and represent a broad range of physicochemical properties (i.e., pI, molecular weight). These are therefore ideally suited as model proteins, which are primarily used to demonstrate the performance and suitability of ultrafiltration for protein fractionation, and for some more esoteric theoretical studies.

In Table 12.7, an attempt has been made to summarise the reports on the application of ultrafiltration for purification of proteins from complex mixtures.

Table 12.7. Fractionation of proteins from complex mixtures using ultrafiltration.

Proteins	Membrane(s)	Parameters examined	Other comment/ Observations	Reference
Lysozyme from egg white	25 and 50 kDa polysulfone	P J H	Two-step process examined	[47]
Lysozyme from egg white	50 kDa polysulfone (surface modified)	P J H M	Membrane pre-treated with protein	[46]
Lysozyme from egg white	30 kDa polysulfone membrane	P J H C	Hollow fibre membrane	[17]
Hemoglobin and myoglobin	30 and 150 kDa inorganic membrane	P J S		[24]
β-Xylanases from media	6, 20 and 50 kDa polysulfone	P	Zeta potential studied	[10]
Ovalbumin from egg white	50 kDa polysulfone	M P S		[11]
Clam viscera proteinases	30 kDa polysulfone	H J		[41]
Serum cytochrome C	Different membranes			[42]
Protein hydrolysates	Membranes with MWCO between 20 and 40 kDa			[43]

Table 12.7. *Continued.*

Proteins	Membrane(s)	Parameters examined	Other comment/ Observations	Reference
Soy protein hydrolysates	5, 10 and 50 kDa membranes			[44]
Whey proteins	100 kDa membrane	P S		[25]
Whey proteins	Different membranes	P	Effect of calcium chelation examined	[26]
Plasma proteins	Cellulose diacetate membrane		Effects of protein adsorption examined	[21]
Whey β-Casein monomer	Carbosep M1, M9 and M6 membranes			[45]
HSA and human IgM	PF100 Cellulose diacetate and Eval2A hollow fibre membranes		Effect of pressure and time examined	[50]
HSA and human IgM	PF100 Cellulose diacetate hollow fibre membrane	J H	Effect of pressure and time examined	[51]
Bovine albumin from plasma	PF100 Cellulose diacetate hollow fibre membrane		Effect of pressure and cascade filtration examined	[52]
Human albumin from plasma	PF100 Cellulose diacetate hollow fibre membrane		Effect of pressure and cascade filtration examined	[52]
Human albumin from plasma	PF100 Cellulose diacetate hollow fibre membrane	J H	Effect of pulsatile flow examined	[53]
HSA and human IgM	PF100 Cellulose diacetate hollow fibre membrane	J H	Effect of pulsatile flow examined	[53]

Parameters:
M = Membrane surface modification, H = System hydrodynamics,
P = pH, J = Permeate flux, S = Salt concentration,
C = Protein concentration

The potential for use of ultrafiltration for protein fractionation is in two main areas:

a. Fractionation of bulk proteins
b. Fractionation of high value proteins

12.5.1   *Fractionation of bulk proteins*

This represents the high volume, low value application segment. Proteins in this category include certain high volume blood products like albumin, and food and dairy proteins. Proteins commonly fractionated in large quantities include serum albumin, polyclonal immunoglobulins, casein, alpha lactalbumin, beta lactoglobulin, ovalbumin and industrial enzymes. These are mostly fractionated using conventional methods. This is one area where ultrafiltration is likely to find widespread acceptance. The high throughput of ultrafiltration and ease of scale-up are the main incentives for the use of ultrafiltration in this area.

12.5.2   *Fractionation of high value proteins*

This represents the low bulk, high value application segment. Here again, the high throughput and ease of scale-up of ultrafiltration are likely to prove advantageous, since most biotechnological product streams are dilute with respect to the product and large quantities of fluids have to be processed to get reasonable amounts of target molecules. If the product is in the permeate, another major advantage of using ultrafiltration is that the product is free from foreign particles. If such an operation is carried out under sterile conditions, pure microbial contamination free product streams (suitable for pharmaceutical use) can be obtained. Candidate proteins for purification by ultrafiltration include monoclonal antibodies, blood proteins such as Factor VIII, urokinase and t-PA.

References

1. K.C. Ingham, T.F. Busby, Y. Sahlestrom and F. Castino, 'Separation of macromolecules by ultrafiltration: Influence of protein adsorption, protein–protein interactions and concentration polarisation', A.R. Cooper (ed.)

Polymer Science and Technology: Ultrafiltration Membranes and Applications Vol. 13, Plenum Press, New York (1980) p. 141.

2. S. Ohno, K. Koyama and M. Fukuda, *US Patent 4*, 347, 138 (1981).
3. S. Nakao *et al.*, 'Separation of proteins by charged ultrafiltration membranes' *Desalination* **70** (1988) 191.
4. A. Higuchi, S. Mishima and T. Nakagawa, 'Separation of proteins by surface modified polysulfone membranes' *Journal of Membrane Science* **57** (1991) 175.
5. N.N. Sudareva, O.I. Kurenbin and B.G. Belenkii, 'Increase in the efficiency of membrane fractionation' *Journal of Membrane Science* **68** (1992) 263.
6. S. Saksena and A.L. Zydney, 'Effect of solution pH and ionic strength on the separation of albumin from immunoglobulins (IgG) by selective filtration' *Biotechnology and Bioengineering* **43** (1994) 960.
7. E. Iritani, Y. Mukai and T. Murase, 'Upward dead-end ultrafiltration of binary protein mixtures' *Separation Science and Technology* **30** (1995) 369.
8. R.H.C.M van Eijndhoven, S. Saksena and A.L. Zydney, 'Protein fractionation using electrostatic interactions in membrane filtration' *Biotechnology and Bioengineering* **48** (1995) 406.
9. M. Balakrishnan and G.P. Agarwal, 'Protein fractionation in a vortex flow filter. II: Separation of simulated mixtures' *Journal of Membrane Science* **112** (1996) 75.
10. N. Ehsani, M. Nyström, H. Ojamo and M. Siikaaho, 'Separation of enzymes produced by Trichoderma reesei with hydrophobic ultrafiltration membranes' *Process Biochemistry* **31** (1996) 253.
11. N. Ehsani, S. Parkkinen and M. Nyström, 'Fractionation of natural and model-egg white protein solutions with modified and unmodified polysulfone UF membranes' *Journal of Membrane Science* **123** (1997) 105.
12. N.P. Perepechkina and L.P. Perepechkin, 'Efficient molecular mass fractionation of leukocyte extract by membrane separation' *Journal of Membrane Science* **160** (1996) 1.
13. Q.Y. Li, Z.F. Cui and D.S. Pepper, 'Fractionation of HSA/IgG by gas sparged ultrafiltration' *Journal of Membrane Science* **136** (1997) 181
14. R. Ghosh and Z.F. Cui, 'Fractionation of BSA and lysozyme using ultrafiltration: Effect of pH and membrane pretreatment' *Journal of Membrane Science* **139** (1998) 17.
15. R. Ghosh, Q.Y. Li and Z.F. Cui, 'Fractionation of BSA and lysozyme using ultrafiltration: Effect of gas sparging' *AIChE Journal* **44** (1998) 61.
16. Q.Y. Li *et al.*, 'Enhancement of ultrafiltration by gas sparging with flat sheet membrane modules' *Separation and Purification Technology* **14** (1998) 79.
17. R. Ghosh, S.S. Silva and Z.F. Cui, 'Lysozyme separation by hollow fibre ultrafiltration' *Biochemical Engineering Journal* **6** (2000) 19.

18. R. Ghosh, 'Fractionation of biological macromolecules using carrier phase ultrafiltration' *Biotechnology and Bioengineering* **74** (2001) 1.

19. H. Miyama, K. Tanaka, Y. Nosaka and N. Fujii, 'Charged ultrafiltration membranes for permeation of proteins' *Journal of Applied Polymer Science* **36** (1988) 925.

20. L. Millesime, J. Dulieu and B. Chaufer, 'Fractionation of proteins with modified membranes' *Bioseparation* **6** (1996) 135.

21. C. Legallais, M.Y. Jaffrin and J. Wojcicki, 'The effect of protein adsorption on APO-B removal and selectivity in plasma fractionation by membrane' *Journal of Membrane Science* **91** (1994) 97.

22. L. Zhang and H.G. Spencer, 'Selective separation of proteins by microfiltration with formed-in-place membranes' *Desalination* **90** (1993) 137.

23. S. Nakatsuka and A.S. Michaels, 'Transport and separation of proteins by ultrafiltration through sorptive and non-sorptive membranes' *Journal of Membrane Science* **69** (1992) 189.

24. F. Sannier *et al.*, 'Separation of hemoglobin and myoglobin from yellowfin tuna red muscle by ultrafiltration: Effect of pH and ionic strength' *Biotechnology and Bioengineering* **52** (1996) 501.

25. R.K. Mehra and W.J. Donnelly, 'Fractionation of whey-protein components through a large pore-size, hydrophilic, cellulosic membrane' *Journal of Dairy Research* **60** (1993) 89.

26. M. Kawasaki *et al.*, 'pH dependent molecular weight changes of kappa-casein glycomacropeptide and its preparation by ultrafiltration' *Milchwissenschaft — Milk Science International* **48** (1993) 191.

27. F. Nau, F.L. Kerherve, J. Leonil and G. Daufin, 'Selective separation of tryptic beta-casein peptides through ultrafiltration membranes — Influence of ionic interactions' *Biotechnology and Bioengineering* **46** (1995) 246.

28. C.S. Slater, T.G.Jr Huggins, C.A.III Brooks and H.C. Hollein, 'Purification of alkaline phosphatase by ultrafiltration in a stirred batch cell' *Separation Science and Technology* **21** (1986) 575.

29. M. Krishnan, N. Kalogerakis, L.A. Behie and A.K. Mehrotra, 'Separation of monoclonal IgM antibodies using tangential flow ultrafiltration' *Canadian Journal of Chemical Engineering* **72** (1994) 982.

30. S. Najarian and B.J. Bellhouse, 'Effect of liquid pulsation on protein fractionation using ultrafiltration processes' *Journal of Membrane Science* **114** (1996) 245.

31. A. Muller, G. Daufin and B. Chaufer, 'Ultrafiltration modes of operation for separation of α-lactalbumin from acid casein whey' *Journal of Membrane Science* **153** (1999) 9.

32. R. van Reis *et al.*, 'High-performance tangential flow filtration using charged membranes' *Journal of Membrane Science* **159** (1999) 133.

33. R. van Reis *et al.*, 'High performance tangential flow filtration' *Biotechnology and Bioengineering* **56** (1997) 71.

34. R. van Reis *et al.*, 'Linear scale ultrafiltration' *Biotechnology and Bioengineering* **55** (1997) 737.

35. R. van Reis *et al.*, 'Constant C wall ultrafiltration process control' *Journal of Membrane Science* **130** (1997) 123.

36. R.G. Nel, S.F. Oppennheim and V.G.J. Rodgers, 'Effects of solution properties on solute permeate flux in bovine serum albumin — IgG ultrafiltration' *Biotechnology Progress* **19** (1994) 960.

37. M. Nystrom *et al.*, 'Fractionation of model proteins using their physicochemical properties' *Colloids and Surfaces A: Physicochemical and Engineering Aspects* **138** (1998) 185.

38. R. Shukla, M. Balakrishnan and G.P. Agarwal, 'Bovine serum albumin-hemoglobin fractionation: Significance of ultrafiltration system and feed solution characteristics' *Bioseparation* **9** (2000) 7.

39. K. Sakhamuru, D.H. Hough and J.B. Chaudhuri, 'Protein purification by ultrafiltration using a β-Galactosidase fusion tag' *Biotechnology Progress* **16** (2000) 296.

40. R. Schnabel, P. Langer and S. Breitenbach, 'Separation of protein mixtures by Bioran® porous glass membranes' *Journal of Membrane Science* **36** (1988) 55.

41. H.C. Chen and R.R. Zall, 'Concentration and fractionation of clam viscera proteinases by ultrafiltration' *Process Biochemistry* **20** (1985) 46.

42. M.D. Goncalves and F. Galenbeck, 'Serum protein fractionation by membrane processes — Centrifugal ultrafiltration, osmosedimentation and multistage ultrafiltration' *Separation Science and Technology* **24** (1989) 303.

43. R. Audinos and J.L. Branger, 'Ultrafiltration concentration of enzyme hydrolysates' *Journal of Membrane Science* **68** (1992) 195.

44. W.D. Deeslie and M. Cheryan, 'Fractionation of soy protein hydrolysates using ultrafiltration membranes' *Journal of Food Science* **57** (1992) 411.

45. O. Lebere and G. Daufin, 'Fouling and selectivity of membranes during separation of beta-casein' *Journal of Membrane Science* **88** (1994) 263.

46. R. Ghosh and Z.F. Cui, 'Protein purification by ultrafiltration with pre-treated membrane' *Journal of Membrane Science* **167** (2000) 47.

47. R. Ghosh and Z.F. Cui, 'Purification of lysozyme using ultrafiltration' *Biotechnology and Bioengineering* **68** (2000) 191.

48. M. Balakrishnan and G.P. Agarwal, 'Protein fractionation in a vortex flow filter. I: Effect of system hydrodynamics and solution environment on single protein transmission' *Journal of Membrane Science* **112** (1996) 47.

49. R. Ghosh and Z.F. Cui, 'Simulation study of the fractionation of proteins using ultrafiltration' *Journal of Membrane Science* **180** (2000) 29.

50. C. Charcosset, M.Y. Jaffrin and L.-H. Ding, 'Time and pressure dependence of sieving coefficients during membrane plasma fractionation' *ASAIO Transactions* 36 (1990) M594.
51. C. Charcosset, M.Y. Jaffrin and L.-H. Ding, 'Effect of permeate to feed ratio on protein recovery in membrane plasma fractionation' *Journal of Membrane Science* **60** (1991) 87.
52. C. Charcosset, L.-H. Ding, M.Y. Jaffrin and U. Baurmeister, 'Comparison of albumin recovery in cascade filtration for human and bovine plasma' *Artificial Organs* **14** (1990) 219.
53. L.-H. Ding, C. Charcosset and M.Y. Jaffrin, 'Albumin recovery enhancement in membrane plasma fractionation using pulsatile flow' *The International Journal of Artificial Organs* **14** (1991) 61.

Chapter 13

New Developments

13.1 New membranes

This is an area where membrane manufacturers and membrane development groups are putting in a lot of effort. A list of membrane manufacturers is given in Appendix B. There is now a much better understanding of the membrane chemistry and the chemistry of the separation processes. A large amount of membrane related information is available on the world-wide web. Appendix C lists some of the membrane related websites.

New membranes are being developed, based mainly on the following considerations:

a. Applicability over wider pH range
b. Applicability with wider range of solvents
c. Smaller pore size distribution
d. Added mechanical strength
e. Lower protein adsorption and fouling
f. Desired levels of hydrophobicity/hydrophilicity

Most of the research on the development of novel membranes is focussed on reduction of protein adsorption. Several of these novel membranes are prepared by chemical modification of membranes already being used. Pieracci *et al.* [1] have discussed two different techniques by which polyethersulfone membrane can be photochemically modified with the monomer N-vinyl-2-pyrrolidinone. This modification increases the wettability of the membrane and decreases adsorptive fouling. These improvements were demonstrated by ultrafiltering BSA solution. Plasma induced immobilisation of PEG on PVDF membrane was investigated as a means by which protein adsorption could be reduced [2]. The PEG was coated on the membrane surface, followed by dipping of the membrane in a PEG/CHCl$_3$ solution and subsequent plasma exposure. A PEG graft concentration ratio above 3.2 was found to induce good fouling resistant properties. Wang *et al.* [3] have reported the synthesis of a copolymer membrane with anti-fouling properties. The membrane was prepared from the P(PEGMA)-graft-PVDF copolymer by a phase-inversion method. The fouling resistant property of this membrane was studied using BSA. It was found that P(PEGMA)-graft-PVDF copolymer with a graft concentration ratio (i.e., [PEGMA]/[CH2CF2] ratio) above 0.06 was effective in preventing protein adsorption. The preparation of a novel ultrafiltration membrane using a polyacrylonitrile co-polymer has recently been reported [4]. This membrane was found to be hydrophilic and thus fouling resistant. The use of heat treatment for the preparation of a hydrophilic PVA membrane with good mechanical properties has been reported by Amanda and Mallapragada [5]. This membrane was also found to be fouling resistant.

13.2   New membrane modules

Membrane module design plays a very large role in determining the efficiency of a process, since it influences the system hydrodynamics and transmembrane pressure. Also, conventional modules may not be suitable for newer applications of membrane technology. Simple design alterations may sometimes be made to existing membrane modules to enhance efficiency (e.g. gas sparging in tangential flow modules). Sometimes newer designs mean more radical changes from existing configurations. Examples of these include vortex flow ultrafiltration modules, spiral tube geometry and rotating membrane disks.

Protein ultrafiltration using vortex flow devices has been reported [6, 7]. In such devices, the transmembrane pressure and the cross-flow rate can be independently adjusted. Such de-coupling of transmembrane pressure from cross-flow rate has several advantages, particularly in protein fractionation processes. However, such devices consume a lot of energy and hence are suitable only for very high value products. Another concern is the generation of heat, particularly at high rotation speeds. This problem can be overcome by providing cooling arrangements within the device or by pre-chilling the feed. Again, this involves additional costs, thus effectively limiting the use of these devices for processing expensive products.

As mentioned earlier, gas sparging has been found to be useful for increasing the efficiency of ultrafiltration processes. Gas-sparged ultrafiltration can be carried out using conventional ultrafiltration modules, e.g. tubular, hollow fibre. However, tubular membrane modules are generally considered to be more suitable than other types of membrane modules for gas-sparged processes [8]. In a recent paper [9], the characteristics of gas-liquid two-phase flow in small diameter inclined tubes has been discussed. More recently [10], the effects of tubular module inclination on ultrafiltration efficiency have been examined.

13.3 Rapid process optimisation techniques

Protein fractionation using ultrafiltration is feasible. However, a considerable fine-tuning exercise is generally needed for determining conditions optimal for fractionation. Operating parameters that need to be optimised include pH, salt concentration, feed concentration, permeate flux and module hydrodynamics. This involves extensive experimentation, which is time consuming and results in consumption of large amounts of pure proteins. Some of these substances are very expensive (e.g. urokinase, streptokinase, t-PA, monoclonal antibodies). A novel experimental technique for determining the sieving coefficients of different proteins with respect to different ultrafiltration membranes has been discussed by Ghosh and Cui [11]. This technique is based on constant J_v ultrafiltration in a stirred cell module, which is integrated with a standard liquid chromatography system (see Fig. 13.1). A small sample (pure protein solution) is injected into the module in the form of a pulse, and the protein concentration in the permeate is continuously monitored using an online flow-through UV detector. The instantaneous protein concentrations in the bulk feed and on the membrane surface are determined using an integral equation based on a 'quasi steady state' assumption. This technique facilitates rapid determination of optimal fractionation conditions with minimum consumption of pure proteins. Recently, the scope of this technique has been extended to study fouling of membranes by protein [12].

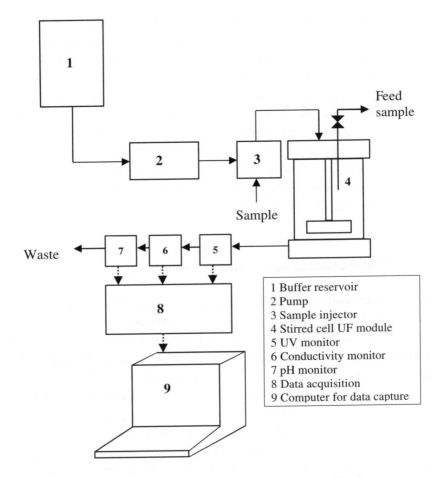

Fig. 13.1. Experimental set-up for pulsed injection ultrafiltration.

13.4 New modes of operation

The mode of operation has a significant influence on performance. Based on the flow behavior of feed with respect to the membrane, ultrafiltration can be classified into dead-end and cross-flow modes. Based on the manner in which feed, retentate and product streams are handled, ultrafiltration processes can broadly be classified into four different modes (see Fig. 4.13). Most industrial operations use a combination of these different modes, depending on specific requirements. In a recent paper,

Muller *et al.* [13] discussed the effect of three different modes of operation (i.e., continuous concentration, discontinuous concentration and diafiltration) on purity and yield of product in the separation of alpha lactalbumin from acid casein whey. Some of the recent developments in the area of novel modes of operation are:

13.4.1 *High Performance Tangential Flow Filtration (HPTFF)*

Ultrafiltration performances can be very significantly improved by using HPTFF [14–16]. This involves the use of two-stage closed-loop cascade separation. In HPTFF, membrane fouling is minimised by operating the ultrafiltration process in the transmembrane pressure-dependent regime of the filtrate flux curve. Fouling is further minimised by controlling the fluid dynamic start-up conditions. Concentration polarisation is generally regarded as a limiting factor in ultrafiltration processes. However, in HPTFF, concentration polarisation is exploited to enhance resolution of solutes. A co-current filtrate stream (similar to a sweep stream) controls concentration polarisation. This arrangement maintains the transmembrane pressure constant along the length of the membrane module. BSA and IgG could be separated very efficiently using this system [14].

13.4.2 *Linear Scale ultrafiltration*

Linear scale ultrafiltration is a new design methodology, which allows a 400-fold linear scale-up from laboratory scale without intermediate pilot scale tests [17]. This is achieved by maintaining constant fluid dynamic parameters, which are independent of scale. Conventional ultrafiltration processes are generally scaled up by keeping the permeate flux, the feed side pressure and the filtrate side pressure constant. It has been argued in this paper that predictable scale up can only be achieved by maintaining constant fluid dynamic parameters, which are independent of scale. Fluid dynamics within a membrane module depend on the feed flow rate, module geometry and special arrangements, such as turbulence promoters.

13.4.3   *Constant C wall ultrafiltration*

Most ultrafiltration processes are operated at constant transmembrane pressure. An alternative strategy is to carry out the process by maintaining a constant wall concentration (C_w) of retained protein [18]. Protein sieving coefficients, selectivity and adsorption on membranes are dependent on C_w. It is expected that by operating at constant wall concentration, a much greater degree of control over the process might be possible.

13.4.4   *Carrier phase ultrafiltration (CPUF)*

Some of the problems associated with conventional ultrafiltration processes are:

a. Most bioprocess feed streams contain different buffering species (e.g. ions, proteins) and it is difficult to adjust the pH of these solutions/suspensions without addition of concentrated acid or alkali, which cause severe denaturation

b. Due to the presence of different ionic substances, it is difficult to precisely adjust the ionic strength

c. Most ultrafiltration processes are operated at constant pressure. This results in a decrease in permeate flux during the process leading to change in protein transmission. Thus, the selectivity of separation can vary with time

d. In cross-flow filtration, the concentration polarisation increases along the length of the membrane. For high-resolution purification, uniform concentration polarisation at all points on the membrane might be desirable

e. Change in pressure drop along the length of the membrane is observed in cross-flow filtration. This can be corrected by counter-current flow of permeate, but this introduced complications in terms of equipment design and operation

f. For protein fractionationation, maintaining constant J_v and k values is desirable. J_v is difficult to maintain constant in conventional filtration

g. Tangential flow filtration is a multi-pass operation in which the proteins will pass through pumps and valves several times. Protein degradation is likely to take place within these system components

h. In tangential flow filtration, permeate flux and mass transfer coefficient cannot be de-coupled. This imposes limitations from the point of view of separation efficiency

Some of these problems can be overcome by using a novel mode of operation termed Carrier Phase Ultrafiltration (CPUF) [19]. CPUF is based on a modification of dead-end UF. A carried phase, which is a buffer corresponding to the optimised physicochemical conditions (i.e., buffer type, pH and ionic strength), is filtered through a UF unit in which the permeate flux and system hydrodynamics can be adjusted independently (e.g. by using a stirred cell or a vortex flow UF unit). The flow rate through the membrane (corresponding to the optimised permeate flux value) is maintained constant by pumping the carrier phase into the module using a high performance pump (e.g. gear pump, plunger pump, needle pump). CPUF is carried out at constant flux and hence the pressure can change during the operation. The optimised hydrodynamic conditions (i.e., values of k) are maintained using suitable stirring or rotating arrangements (e.g. magnetic or shaft driven stirrer, rotating cylinder). All these ensure that UF can be carried out under optimised conditions throughout an entire operation cycle.

Some of the more obvious advantages of CPUF are:

a. There is no need to adjust pH and salt concentration of the feed
b. By using a high performance pump which can provide a constant flow rate even at very high backpressure, the process can be operated at a fixed permeate flux (J_v)
c. By selection of appropriate membrane modules, the process can be carried out at relatively constant mass transfer coefficient (k) at all points on the membrane
d. In CPUF, the protein molecules go through pumps and valves only once. Therefore damage is minimised
e. In CPUF, the permeate flux and the mass transfer coefficient can be de-coupled
f. It is possible to have a greater degree of control over the process

References

1. J. Pieracci, D.W. Wood, J.V. Crivello and G. Belfort, 'UV-assisted graft polymerisation of N-vinyl-2-pyrrolidinone onto poly(ether sulfone) ultrafiltration membranes: Comparison of dip versus immersion modification techniques' *Chemistry of Materials* **12** (2000) 2123.
2. P. Wang, K.L. Tan, E.T. Kang and K.G. Neoh, 'Plasma-induced immobilization of poly(ethylene glycol) onto poly(vinylidene fluoride) microporous membrane' *Journal of Membrane Science* **195** (2002) 103.
3. P. Wang, K.L. Tan, E.T. Kang and K.G. Neoh, 'Synthesis, characterisation and anti-fouling properties of poly(ethylene glycol) grafted poly(vinylidene fluoride) copolymer membranes' *Journal of Materials Chemistry* **11** (2001) 783.
4. H. Espinoza-Gomez and S.W. Lin, 'Development of ultrafiltration membranes from acrylonitrile co-polymers' *Polymer Bulletin* **47** (2001) 297.
5. A. Amanda and S.K. Mallapragada, 'Comparison of protein fouling on heat-treated poly(vinyl alcohol), poly(ether sulfone) and regenerated cellulose membranes using diffuse reflectance infrared Fourier transform spectroscopy' *Biotechnology Progress* **17** (2001) 917.
6. R. Shukla, M. Balakrishnan and G.P. Agarwal, 'Bovine serum albumin-hemoglobin fractionation: Significance of ultrafiltration system and feed solution characteristics' *Bioseparation* **9** (2000) 7.
7. M. Balakrishnan and G.P. Agarwal, 'Protein fractionation in a vortex flow filter 2. Separation of simulated mixtures' *Journal of Membrane Science* **112** (1996) 75.
8. R. Ghosh and Z.F. Cui, 'Mass transfer in gas-sparged ultrafiltration: Upward slug flow in tubular membranes' *Journal of Membrane Science* **162** (1999) 91.
9. T.W. Cheng and T.L. Lin, 'Characteristics of gas-liquid two-phase flow in small diameter inclined tubes' *Chemical Engineering Science* **56** (2001) 6393.
10. T.W. Cheng, 'Influence of inclination on gas-sparged cross-flow ultrafiltration through an inorganic tubular membrane' *Journal of Membrane Science* **103** (2002) 103.
11. R. Ghosh and Z.F. Cui, 'Analysis of protein transport and polarisation through membranes using pulsed sample injection technique' *Journal of Membrane Science* **175** (2000) 75.
12. R. Ghosh, 'Study of membrane fouling by BSA using pulsed injection technique' *Journal of Membrane Science* **195** (2002) 117.
13. A. Muller, G. Daufin and B. Chaufer, 'Ultrafiltration modes of operation for separation of α-lactalbumin from acid casein whey' *Journal of Membrane*

Science **153** (1999) 9.

14. R. van Reis *et al.*, 'High performance tangential flow filtration' *Biotechnology and Bioengineering* **56** (1997) 71.
15. R. van Reis *et al.*, 'High-performance tangential flow filtration using charged membranes' *Journal of Membrane Science* **159** (1999) 133.
16. S. Gadam *et al.*, 'Liquid porosimetry technique for correlating intrinsic protein sieving: Applications in ultrafiltration processes' *Journal of Membrane Science* **133** (1997) 111.
17. R. van Reis *et al.*, 'Linear scale ultrafiltration' *Biotechnology and Bioengineering* **55** (1997) 737.
18. R. van Reis *et al.*, 'Constant C wall ultrafiltration process control' *Journal of Membrane Science* **130** (1997) 123.
19. R. Ghosh, 'Fractionation of biological macromolecules using carrier phase ultrafiltration' *Biotechnology and Bioengineering* **74** (2001) 1.

Glossary of Membrane Terminology

A diverse array of terms and definitions are used in membrane research and applications. In order to ensure some degree of uniformity in membrane terminology, the International Union of Pure and Applied Chemistry (IUPAC) set up a *Working Party on Membrane Nomenclature* to recommend terminology for membranes and membrane processes. A link to this is available at the North American Membrane Society website [www.membranes.org].

Some of the terms and definitions commonly used in membrane research and practice are given below:

Apparent rejection coefficient (R_a): This equals one minus the ratio of concentrations of a component in the permeate stream and in the bulk feed

Apparent sieving coefficient (S_a): This is the ratio of the concentrations of a component in the permeate stream and in the bulk feed

Asymmetric $(anisotropic)$ *membrane*: A composite membrane, consisting of two or more structural planes of non-identical morphologies

Backflush: Temporary reversal of the direction of the permeate flow, used mainly to un-block membrane pores

Boundary layer: Fluid layer in contact with solid surface, where the flow velocity is lower than in the bulk fluid

Bubble point: Transmembrane pressure, at which bubbles first appear on the downstream surface of a porous membrane immersed in an appropriate liquid as pressure is applied to the upstream side

Buffer exchange: Change of buffering species by diafiltration

Cake layer: A layer made up of rejected and subsequently deposited particles on the upstream face of a membrane

Cascade ultrafiltration: Multi-stage ultrafiltration, analogous to multi-stage distillation

Clarification: Removal of particles

Co-current flow: Flow pattern in a membrane module, whereby the fluids on both sides of the membrane move parallel to the membrane surface in the same direction

Completely mixed (perfectly mixed) flow: Flow pattern in a membrane module, in which fluids on both sides of the membrane are individually well mixed

Composite membrane: Membrane made up of chemically or physically distinct layers (e.g. asymmetric membrane)

Concentration (process): Removal of solvent from a solution to increase the solute concentration

Concentration polarisation: Accumulation of higher levels of solute close to the membrane surface in comparison to the well-mixed bulk feed

Concentration factor: Ratio of the concentration of a component in the retentate to that in the feed

Continuous membrane column: Membrane modules arranged analogous to that of stages in a distillation column (also called *cascade membrane filtration*)

Counter-current flow: Flow pattern in a membrane module, in which the fluids on both sides of the membrane move parallel to the membrane surface but in opposite directions

Critical flux: Permeate flux below which fouling is not severe

Cross-flow: Flow pattern in a membrane module, in which the fluid on the upstream side moves parallel to the membrane surface while the fluid on the downstream side moves away in a direction normal to the membrane surface

Dead-end flow: Flow pattern in a membrane module, where the feed flows towards the membrane in a normal direction while the permeate flows away from the membrane, also in a normal direction

Dean vortex: Vortex formed due to flow instability in curved flow passages

Dense (non-porous) membrane: Membrane with no detectable pores

Desalting: Removal of salts from protein solutions by diafiltration

Diafiltration: A simple ultrafiltration process, in which low molecular weight contaminating solutes are removed from solutions of biopolymers, such as proteins, accompanied by continuous replacement of solvent lost with the permeate

Diffusivity: A measure of the ability to undergo diffusive transport

Dynamic membrane: An active layer that is formed on the membrane surface by the deposition or accumulation of rejected species

Electrostatic double layer: Layer formed around charged macromolecules and charged surfaces by ionic species present in solution

Electroultrafiltration: Ultrafiltration facilitated by an electric field

Fouling: Deposition or adsorption of particulate or dissolved substances on the external membrane surface, at the pore openings, or within the pores leading to loss of efficiency

Gas-sparging: The method for permeate flux enhancement which employs gas bubbles as turbulence promoters

Gel layer: A gelatinous fouling layer formed by rejected colloidal material on the membrane surface

Hindered transport: Hindered diffusion and convection through a membrane, due to retarding forces such as electrostatic interactions and drag forces

Hollow fibre: A fine tubular membrane

Homogeneous membrane: Membrane with essentially the same structural and transport properties throughout its thickness (also referred to as *isotropic membrane*)

Hydraulic permeability: The water flux divided by applied transmembrane pressure

Intrinsic membrane rejection: Rejection of a solute by a native membrane

Intrinsic rejection coefficient (R_i): This equals one minus the ratio of concentrations of a component in the permeate and that on the wall of the upstream side of the membrane

Intrinsic sieving coefficient (S_i): This is the ratio of the concentrations of a component in the permeate and that on the wall of the upstream side of the membrane

Laminar flow: Fluid flowing slowly and smoothly with the fluid streamlines parallel to each other

Leaching: Removal of specific species from polymer films to produce the membrane pores

Limiting flux: Maximum achievable permeate flux for a given hydrodynamic condition

LRV: Log Removal Value, which is a measure of the efficiency of removal of particulate matter by microporous membranes

Mass transfer coefficient: This is the measure of diffusive velocity of a solute within a region, such as a film, and is obtained by dividing the diffusivity by the diffusive path length

Maxwell-Stefan modelling: Analysis of membrane transport using Maxwell-Stefan equation

Membrane ageing: Change in membrane transport properties over a period of time due to physical, chemical or structural alterations

Membrane casting: Preparation of a membrane using a solution of the membrane-forming polymer

Membrane conditioning (*pretreatment*): Process carried out on a membrane after its preparation, in order to improve its properties

Membrane module (*cell*): Device within which the membrane elements (such as flat sheet and hollow fibre) are housed

Membrane reactor: A reactor employing a membrane either for retention of specific species or for immobilising the catalyst

Microfiltration: A pressure driven membrane based separation process primarily used for the separation of particles from fluids

Molecular-weight cut-off ($MWCO$): Molecular weight of a solute corresponding to a 90 percent rejection coefficient ($R_a = 0.9$) for a given membrane

Nanofiltration: A pressure driven membrane based separation process primarily used for the separation of medium sized molecules

Observed transmission (τ_{ob}): This is the ratio of the concentration of a component in the permeate and that in the feed (same as *apparent sieving coefficient*)

Oscillatory flow: Oscillation of feed flow to minimise concentration polarisation and thus increase permeate flux

Osmotic pressure: Pressure exerted due to difference in solute concentration between two distinct zones

Permeability: This is defined as a transport flux per unit transmembrane driving force

Permeate: Stream containing the species that have penetrated the membrane (also called *filtrate*)

Permeate flux: Volume or mass of liquid crossing the membrane per unit area per unit time

Pore tortuosity: Ratio of actual pore length to membrane thickness

Porosity: Ratio of void space to total membrane volume in porous membranes

Pressure pulsing: The use of periodic negative transmembrane pressure, in order to improve permeate flux in ultrafiltration processes

Real or intrinsic transmission (τ): This is the ratio of the concentrations of a component in the permeate and that on the wall of the upstream side of the membrane (same as *intrinsic sieving coefficient*)

Relative recovery (η): The amount of a component collected in a useful product stream divided by the amount of that component entering the process

Retentate: The stream leaving the feed side of a cross-flow membrane module or the material left within the feed side in the case of dead-end filtration

Reverse osmosis: A pressure driven membrane based process primarily used for the separation of small molecules

Selectivity (ψ): This represents the efficiency of solute separation, which in binary solute fractionation equals the ratio of observed transmissions for the two solutes

Self-rejection: Rejection of solute due to electrostatic repulsion by membrane bound solute molecules

Skin layer: A thin layer which is located at the upstream face of an asymmetric membrane and which is primarily responsible for determining selectivity and permeability

Sieving coefficient: A measure of the transport of a specific solute through a membrane, usually obtained by dividing the concentration in the permeate by a characteristic upstream concentration (e.g. bulk or wall)

Solute flux: Mass or number of moles of solute crossing the membrane per unit area per unit time

Streaming potential: A transmembrane electrical potential, which arises from the coupling between flux of charged species and solvent through the pores

Tangential flow: Flow of feed parallel to the membrane surface (similar to *cross-flow*)

Taylor vortex: Vortex formed in the annulus of two concentric rotating cylinders

Transmembrane pressure (ΔP): The applied driving force for convective material transport through a membrane (average pressure difference between the upstream and downstream sides of the membrane)

Tube inserts: Turbulence promoters introduced within tubular membranes to improve efficiency of filtration

Turbulent flow: Fluid flow comprising eddies and whorls (vortices), usually observed at high flow rates

Ultrafiltration: A pressure driven membrane based separation process primarily used for the separation of macromolecular substances

Volumetric permeate flux: The volume of permeate collected per unit time per unit membrane surface area

Wall shear rate: This is the liquid shear rate at the membrane surface

Zeta potential: An electrical potential, which is a measure of surface charge

Appendix B

List of Membrane Manufacturers

A/G Technology Corporation
www.agtech.com

Asahi Kasei Corporation
www.asahi-kasei.co.jp/asahi

CeraMem Corporation
www.ceramem.com

Danish Separation System AS
www.dksepsys.dk

Flowgen Instruments Ltd UK
www.flowgen.co.uk

Gelman Science Ltd
www.gelman.com

Hydranautics
www.membranes.com

Koch Membrane Systems Inc
www.kochmembrane.com

Membraflow Filtersysteme GMbH
www.membraflow.de

Membrana
www.membrana.com

Millipore Corporation
www.millipore.com

Nadir Filtration GMbH
www.nadir-filtration.de

Niro Inc
www.niroinc.com

Nitto Denko Corporation
www.nitto.com

Osmonics Inc
www.osmonics.com

Pall Corporation
www.pall.com

PCI Membrane Systems Inc
www.pcims.com

Rhodia GMbH
www.rhodia.com

Sartorius AG
www.sartorius.com

Schleicher & Schuell
www.s-und-s.de

SpinTek Filtration
www.spintek.com

Ultrafilter Inc
www.ultrafilter.com

USF Filtration and Separations
www.usffiltration.com

USFilter Corporation
www.usfilter.com

Appendix C

Websites of Membrane Related Academic and Research Institutions

Centre for Membrane Applied Science and Technology
University of Colorado
www.Colorado.EDU/che/mast

Chemical and Biological Process Engineering Department
University of Wales Swansea
www.engineering.swan.ac.uk/chem_eng.htm

Chemical Engineering Department
University of Bath
www.bath.ac.uk/chem-eng

Chemical Engineering Department
University of Delaware
www.che.udel.edu

Chemical Engineering Department
Rensselaer Polytechnic Institute
www.rpi.edu/dept/chem-eng

Chemical Engineering and Materials Science Department
University of Minnesota
www.cems.umn.edu

Chemical and Process Engineering Research Summary
Oxford University
www.eng.ox.ac.uk/World/Research/Summary/A-Chemical.html

Department of Chemical Engineering
McMaster University
www.chemeng.mcmaster.ca

Desalination
www.elsevier.com/inca/publications/store/5/0/2/6/8/3/index.htt

European Membrane Institute Twente
membrane.ct.utwente.nl

European Membrane Society
www.ems.cict.fr

Industrial Membrane Research Institute
University of Ottawa
www.genie.uottawa.ca/orgs/imri

Institute for Technical Chemistry
University of Essen
www.uni-essen.de/tech2chem/indexen.htm

Journal of Membrane Science
www.elsevier.nl/inca/publications/store/5/0/2/6/9/2

Laboratory of Membrane Technology and Technical Polymer Chemistry
Lappeenranta University of Technology
info.lut.fi/kete/ltpc/LTPC.HTM

McMaster Membrane Research Group
McMaster University
www.chemistry.mcmaster.ca/membrane/

Membrane Online
www.membraneonline.com

North American Membrane Society
www.membranes.org

Swedish Foundation for Membrane Technology
www.membfound.lth.se

UNESCO Centre for Membrane Science and Technology
www.membrane.unsw.edu.au